Dad, Where Did We Come From?
It Was a Dark and Stormy Night!

The Comparison of Cosmic Beginning Accounts

Dr. Marcus O. Durham, Ph.D., Th.D.

Rosemary Durham

Realm Research

Dad, Where Did We Come From?
It Was a Dark and Stormy Night!
The Comparison of Cosmic Beginning Accounts

Contact:
THEWAY Labs
17350 E. US 64 Hwy
Bixby, OK 74008

www.ThewayLabs.com

All figures not original are NASA public domain. NASA material is not protected by copyright unless noted.

Cover Design:
Rosemary Durham & Marcus O. Durham, Ph.D., Th.D.
Photo: NASA Spitzer Space Telescope.
The Serpent Star-Forming Cloud Spawns Stars

Printed in United States of America
First printing, April, 2018

ISBN: 978-1987699937

Copyright © 2017 - 2019 by Marcus O. Durham, Ph.D., Th.D.

All rights reserved under US and International Copyright Law. Contents and/or cover may not be reproduced in whole or in part in any form without the express written consent of the Publisher.

Contents

Chapter 1 5
   The Kids' Question 5
Chapter 2 9
   Sidebar – NASA and Religion 9
Chapter 3 15
   Sequence 15
Chapter 4 23
   Where We Are 23
Chapter 5 27
   NASA 27
Chapter 6 47
   Sidebar - A little math for non-mathematicians 47
Chapter 7 49
   Universal equations 49
      Life cycle 66
Chapter 8 67
   Historical Philosophy 67
      Perspective 78
      A little about ages. 87
Chapter 9 91
   What Other Scientists Say 91
      Current State of Science 91
      NASA Scientist 94
   References 95
Chapter 10 97
   Other Ancient References 97
Chapter 11 101

Unique Acts	101
Chapter 12	**105**
The Final Act	105
Dad, where did we come from?	106
Bibliography	**109**
Annex	**111**
First Law	111
Second Law	112
Third law	112
Dark Matter and Energy	113
About the Authors	**115**
Dr. Marcus O. Durham	115
Rosemary Durham	116
Other Books	117

Chapter 1
The Kids' Question

'It was a dark and stormy night' is the introduction to Snoopy's perpetual novel. The phrase has been perennially used as a euphemism for a melodramatic written novel called purple prose. However, in this case, the phrase is an apt description for real cosmic beginnings.

The beginning of the universe has long been a question for kids to philosophers and scientists. Although the query of 'where did we come from?' is the same, the approach and an acceptable answer is very different.

Our objective is to address the topic on all three levels. The first segment is a straight forward compilation of primarily NASA low-tech information. The second segment is math and science. The third segment is philosophical with a bow to religious tradition.

Our target profile is secondary educated without physics class. Achieving such a goal is challenging, since the topic is steeped in physics and astro-physics. Enough new evidence is presented that information of value should be available across the understanding spectrum.

As a scientist and a theologian, my ideas can run into the deep woods. However, my co-author's daunting task is to drag the information back to understandable reality. Even in the science and math discussion, we hope to use adequate pictures and illustrations so that at least the concepts are visible, if not the subtlety.

Many think they are not interested or gifted in mathematics. That is okay. Every person has different gifts, talents, training, and experience. I am musically challenged. There are issues with my ears that make most music sound like a vibrating cacophony, much like a rattling speaker with a high squeal. Consequently, I seldom listen to the noise. And no one wants to hear my grating, 'ungratiating' singing, not even the shower. So, I seldom sing. Furthermore, with no training, art is a lost art to me. Pun intended. Now science and math to me holds the beauty of the universe.

Fortunately, my co-author is a musician and artist who has an astute eye for color and form. But she does not appreciate the beauty of mathematics. And so it goes with all of us. Consequently, we will try to paint the art of math, which is necessary to explain the subtle, intricate functioning of the universe.

Why should we care about cosmogony and beginnings? That is a fair question. As we will see, understanding the sequence gives the formulas for science, provides the basis for environmental analysis, explains how energy behaves, describes the phenomenon of frequency which has huge health and medical ramifications, explains electro-magnetics which drives all communications, and gives the background into research which we do not even know. In other words, comprehending the beginnings is literally a summation of all knowledge. Your perception of beginnings reveals your world view.

Although we clearly will not cover all knowledge, perhaps we can at least answer the kid question, 'where did we come from?'

The study of origin of the universe or cosmos is called cosmogony from a similar sounding and spelled Greek word. A related word from Latin is genesis. While cosmogony sounds scientific, and genesis sounds religious, the concepts are the same. One of NASA's ventures is the Genesis project, assuredly not a religious investigation.

Major changes have occurred in cosmic understanding during the past 50 years. Although dismissed in the 1920's, still up to 1965, the steady-state or uniformitarian hypothesis persisted in some scientific thinking. Steady-state (uniformitarian) hypothesis assumes the universe was the same as it always had been, so any event simply occurs in the same form as all previous events. In its simplest form, steady state (uniformitarian) is a straight line.

> Until the 1920s, cosmology was still dominated by the theory of a Steady State Universe, or the idea that the universe was homogeneous (has the same general make-up throughout), infinite (that the universe just extends forever), and static (the universe is not expanding, it just is). If you just study the night sky, it seems easy just to think that this is the way the universe has always looked and will always look. In the 20th century, however, observations of the universe did not seem to add up with Steady State theory...
>
> https://lambda.gsfc.NASA.gov/product/suborbit/POLAR/cmb.physics.wisc.edu/tutorial/briefhist.html, recorded 09/27/2017.

Unfortunately, though thoroughly discredited in physics, uniformitarianism in the form of evolutionary thought still permeates biology and the soft sciences of social studies. The persistence of this simplistic process appears to be a religious adherence to tradition because of 'that is what I know' mentality. The consequence is perception that the energy of the universe can be changed from within the system. That analysis conflicts with numerous discernible equations.

Robert Wilson and Arno Penzias, physicists at Bell Laboratories, were working on new antenna designs when they discovered an unusual source of radio noise in 1964. The noise is called cosmic background microwave radiation. The discovery of a persistent noise in the universe pointed to a single incident when the universe was formed. The measurements and calculations show the universe began in less than one-hundredth of one-billionth of one-trillionth of one-trillionth of a second after nothing. That elapsed time may not be zero, but it sure is close.

Called the Big Bang in some circles, the cosmos went from zero to everything instantaneously, in one cataclysmic event. Rather than traditional uniform transitions, hard science realizes all known existing processes are an exponential decay. The rate of decay is determined by the size of repetitive waves or cycles, much like a splash in a pond. For the hard sciences of physics, chemistry, and engineering, the formula is very, very well defined and can be concisely written in one line.

However, the huge ramifications of the single formula are literally out of this world. Much of the social causes would be decimated with an understanding of the consequences of the relationship. Religion would be turned on its ear. Responsibility for actions would be clearly evident. But those are not our objective.

We are simply, concisely identifying the events of the beginning universe. First, we will predominantly use NASA quotes, then formulate the events into concise mathematics and science, finally correlate to ancient historical philosophy. To our knowledge that has not been done in this format. We have researched and written published papers and a couple of books on different aspects of the cosmic processes. This relatively short document is intended to bring these awesome concepts into a simpler formulation.

NASA is generally regarded as a premier scientific organization. The group is best known for its manned space missions. With less emphasis on those high-profile projects, NASA and associated agencies have been quite active in long range probes into space and development of research tools like the Hubbell telescope. The public is only casually aware of these extensive investigations.

NASA's numerous webpages provide an active, massive resource that is generally very readable. Traditional printed books are becoming less of a resource for scientific research. Even traditional scientific journals and technical conference proceedings are giving way to the ubiquitous and available Internet with open-source publication.

Chapter 2
Sidebar – NASA and Religion

Religious tradition is made light of and ridiculed by those of limited understanding. NASA appropriately, rationally, and intellectually addresses the relationship between religion and science.

> *I am religious and I also find science very exciting. Is there a conflict between science and religion?*
>
> *According to the National Academy of Sciences (NAS):*
>
> *"Science is a particular way of knowing about the world. In science, explanations are limited to those based on observations and experiments that can be substantiated by other scientists."*
>
> *"Progress in science consists of the development of better explanations for the causes of natural phenomena. Scientists never can be sure that a given explanation is complete and final. Some of the hypotheses advanced by scientists turn out to be incorrect when tested by further observations or experiments. Yet, many scientific explanations have been so thoroughly tested and confirmed that they are held with great confidence."*
>
> *"Truth in science, however, is never final, and what is accepted as a fact today may be modified or even discarded tomorrow. Science has been greatly successful at explaining natural processes, and this has led not only to increased understanding of the universe but also to major improvements in technology and public health and welfare."*
>
> *The National Academy of Sciences also says:*
>
> *"Science is not the only way of acquiring knowledge about ourselves and the world around us. Humans gain*

understanding in many other ways, such as through literature, the arts, philosophical reflection, and religious experience. Scientific knowledge may enrich aesthetic and moral perceptions, but these subjects extend beyond science's realm, which is to obtain a better understanding of the natural world."

"Scientists, like many others, are touched with awe at the order and complexity of nature. Indeed, many scientists are deeply religious. But science and religion occupy two separate realms of human experience. Demanding that they be combined detracts from the glory of each."

"Many religious persons, including many scientists, hold that God created the universe and the various processes driving physical and biological evolutions and that these processes then resulted in the creation of galaxies, our solar system, and life on Earth. This belief, which sometimes is termed 'theistic evolution,' is not in disagreement with scientific explanations of evolution. Indeed, it reflects the remarkable and inspiring character of the physical universe revealed by cosmology, paleontology, molecular biology, and many other scientific disciplines."

Quotes from: 1999 report "Science and Creationism: A View from the National Academy of Sciences, Second Edition" which is available online from the National Academy Press: http://books.nap.edu/catalog.php?record_id=6024

https://map.gsfc.NASA.gov/site/faq.html, recorded 09/27/2017.

The belief in cosmic development, science, and theology runs the gamut. Three fundamental positions exist, with variations within those philosophies. *Non-causal, philosophical science*, and *theology* are the predominant views. All of us use a mix of these concepts to a varying degree. Three stages divide non-causal as well as theology.

Non-causal implies things happen but the specific cause is undetermined yet. The range of non-causal is randomness, uniformitarian, and hypothetical science.

Randomness assumes events happen but can only be determined by probability. The assumption necessarily comes from incomplete knowledge and incomplete science. German physicist Werner Heisenberg was the principal proponent, which provoked Dr. Albert Einstein to issue his famous declaration, 'God does not throw dice.'

Uniformitarianism assumes the universe has always been the way it is. This static philosophy is the premise of evolution. In other words, one thing leads to another. Interestingly, at the same time that uniformitarian was dismissed as non-credible by the hard science community in the 1920's, evolution became the predominant process of the soft sciences. The hard science equations illustrate that evolution is not viable. The statement is not a religious position and says nothing about the alternatives. Although not intentional, this simple scientific observation challenges accepted political and education agendas and draws the ire of politics. But the scientific method demands reevaluation of accepted theories when new data becomes available. Otherwise there is no progress. Unfortunately, somewhat to placate, the term 'evolution' has slipped into otherwise scientific observation, even by NASA, to describe life and human development. The usage is contrary to NASA's impulse related observations.

Hypothetical science is scientific observations explained by speculative hypothesis. Hypothetical science is the predominant position of cosmic study, and in fact, a significant portion of hard-science. Since the scientific method cannot observe prior to the beginning, then a cause cannot be validated. Unfortunately, dismissal of a design cause leads to many convoluted arguments. I recently watched a presentation about black holes, which have been assumed to be the result of collapsing, dying stars. Now black holes have been found shortly after the beginning of time. Multiple astro-physicists tried to twist their diverse hypotheses to fit the new observations. Repeatedly terms like 'it could be' were interjected. The presentation was actually amusing to this old scientist.

Philosophical science is a middle ground where science of observation is adamantly pursued. Where a cause cannot be found by observation, then philosophy is investigated. Philosophy and theology predate the scientific method by at least 3,000 years of recorded history. Therefore, these studies have information which science can well use.

Theological science is the next step along the philosophical spectrum. The process assumes theological positions are the foundation and can be explained by science. Sense the science is based on observations and hypothesis, this is a valid technique. Leading advocate groups include Answers in Genesis and Creation Research Institute.

The three previous science positions all follow the scientific method and use the same data. A scientist can easily move between the three. The key difference is the basis for hypotheses when there is incomplete data. There is always incomplete data. Incomplete data is a hallmark of cosmic observation and is the thing that has driven astronomers and hard scientists to pursue more information.

Legal theology is the next process. Theology of laws defines traditional organized religion. The theology presumes absolute laws which are established by deity. The observation that some of these laws exist in nature actually borders on science. However, the introduction of other laws, which are not by observation or may be arbitrary separates theology from the scientific method. In general, theology presumes its correctness to the exclusion of science and vice-versa.

Sovereign theology is the final category which assumes a capricious deity. The deity can do anything, including changing or violating its own rules. In some teaching, the deity can be manipulated by religious acts or by supplication. The process creates a negative, victim, fear mentality whereby the end justifies the means. Adherents include radical fundamentalists across religions.

Let's consider how one well-known hard scientist addresses incomplete data. Stephen Hawking, an avowed atheist, stated:

> *I thought I had left the question of the existence of a Supreme Being completely open in my article. It would be perfectly consistent with all we know to say that there was a Being who was responsible for the laws of physics.*
>
> ---Steven Hawking, "The Edge of Space-Time," *American Scientist* 72, 1984, pp 355-359.

But he rejected the God ideas of the latter two theology teachings. In a post-Christian era, how can an extreme theology position be taken seriously by scientists, when the philosophy dismisses the science he knows?

These prejudices determine your world view. Our objective is to present information across the spectrum, but with a leaning to philosophical science.

With these definitions of cosmic analysis, development of scientific theory or philosophical traditions should be possible.

Chapter 3
Sequence

The process of cosmogony is broken into multiple identifiable sequences. The current discussion of this chapter will simply list the sequence with a brief explanation.

Several new words, terms, and concepts may be introduced to you. Be cool, calm, and collected. You can grasp all this at some level. Clearly the physicist and teenager have a different base for learning, but both benefit from having a common thread.

For those of a religious persuasion, please do not panic because you disagree with a particular time frame. In a later section, we will address an alternative discussion, then we will plead with the long-time advocates to make a similar open perspective. The length of time is not critical to our discussion. Rather look at the sequence and the form of the science and equations in juxtaposition with the philosophy, history and religion.

The major challenge to research advancement is the adherence to tradition, whether by traditional religious groups or secular religious groups.

Before you come to a conclusion, have you read the book?

Our hope is to reflect on relatively new information, analyze the science, and identify classical philosophy. The process will raise questions along the journey, but which when resolved, will definitively answer the kids' question.

1. Before the beginning of the cosmos, nothing observable existed.

In other words, the cosmos started from a blank slate.

2. Time begins

In less than one-hundredth of one-billionth of one-trillionth of one-trillionth of a second, the entire cosmos came into existence. The impulse occurred in 1×10^{-34} seconds. That is a decimal followed by 33 zeroes, then a 1. By any definition, that period is instantly.

What happened before time is not observable. Before time is outside any analysis. Before existence is not definable, describable, or comprehendible.

> *Time is simply the measure of elapsed events. The length is an arbitrary period which is agreed usage for how to measure and correlate events. Please note this statement later in time discussions.*

Time may be, but is not necessarily, straight line and one directional.

If all the cosmos components came into being at the instant, then all time must have come into existence. Question to ponder: where is time stored?

Three different times comprise natural science. Fixed time is constant, so it does not change. Cyclic time is repetitive, and is the duration of a single wavelength, cycle, or vibration. Seasonal time has a beginning, duration, and end. This is the linear time most people consider.

3. Matter defined

The condition of going from nothing to a peak value instantaneously is called an impulse or cosmic inflation. The impulse material was unconsolidated (not formed). Matter was a conglomeration of the components of elements. The existing material was an extremely hot morass of particles.

The expansive extremes of cosmic beginnings are so fantastic that the size of the numbers sounds like words a kid would make up to exaggerate. Nonillion degrees! Really?

All matter in the universe today was at the singularity. Matter consists of mass, electric charge, and magnetism.

4. Singularity

The universe was at one point. The point is called a singularity. All future existence was bound at that point. All future time, space, and matter was constrained at the single extremity.

Since something was derived from nothing, clearly the present rules of nature were not in effect. The present rules are contained in the singularity, but the events are undefined yet. The future has not happened.

The cause and effect is defined without the choice of events being proscribed. In other words, if you do this, then something will happen or if you do that, then something else will happen. The path whether to this or that is not established.

The rules of the game are set forth, but the game has not played out. The game results depend on the rules established by the referee, work by players, and choices made by players. In reality, the illustration is the game of life.

The singularity does not involve space.

5. Space, the final frontier

Space did not expand from the point. Rather the very point expanded into space very quickly.

The initial rapid expansion from nothing is called inflation or impulse.

With space defined, only a void or vacuum exists. An old saying affirms 'nature abhors a vacuum'. In other words, nature tries to move material into the void.

The cosmos continues to expand into the previously non-existent space. There is no known or foreseeable boundary or limit on the measurement of space. Therefore, there is no constraint on expansion of the universe. That, ladies and gentlemen, is the definition of infinite, eternal, or everlasting. But, space is not infinite, it is unlimited. There is a subtle but important difference. We just do not know the limit or edge of space.

Two less than perfect analogies illustrate the expansion of material from a small space to fill a larger void. A balloon starts very small but expands into a larger space. Similarly, raisin bread expands from a compact dough when yeast rises to a loaf. As a balloon expands, every point on the surface is moving away from other points.

6. Total darkness

The morass of matter parts existed in total darkness. Darkness is the absence of light. Light is the visible evidence of energy conversion.

Energy is one of those fundamental terms which defies a simple explanation.

Energy is the activity which makes things happen.

It includes work, heat, and situations not observable with the five senses, like electricity.

Aristotle understood that the universe consists of energy and intelligence.

The things we have seen so far are *matter*, *space*, and *time*, which are the things which make energy. Now light as an energy form includes matter which has mass with electro-magnetism and wave frequency. At this point of cosmic beginnings, all the components exist, but have not been consolidated into energy. So, there can be no light yet.

7. Energy equation

Because of the extreme energy resulting from the creative impulse, materials can only exist as a gas. Sparks of light started as the first gaseous elements are fused.

The process called fusion is clashing of components to create more complex material. The opposite process, fission, is breaking apart elements to form smaller, simpler components. The simplest possible fusion combination is one positively charged mass unit and one negatively charged unit. That combination is called hydrogen.

Visible results of the energy beginnings are obtained by measurement of radiation residue, called cosmic microwave background radiation.

All fundamental energy relationships have now been defined in the beginning. *Since that instant nothing can be created or destroyed, but can only change form.* And that is one definition of the First Law of Thermodynamics.

The first epoch is complete!

8. Liquid precipitates

In the second phase of cosmic development, material changes form, but no new fundamental component of nature is involved.

The first sparks of energy result from formation of gases. The most fundamental is hydrogen, followed by helium. Although hydrogen is ubiquitous, helium is second most common. Because helium is heavier, strangely it represents 24% of the total mass of the known universe. However, helium is relatively rare in the earth realm.

At very high temperature, everything is a gas. Vapor or gas coalesced (combined) into liquid. This phase change is an indication and result of cooling.

In a natural system, energy decays from a higher potential to a lower potential, with an unrecoverable loss transferred to the cosmos. The loss is in the form of Entropy multiplied by absolute temperature.

That statement is one definition of the Second Law of Thermodynamics.

Additional materials can fuse at cooling temperatures.

9. Solids coalesce

The third phase of development occurs when liquid is separated from solids. The transition is a further indication of cooling. The same matter is involved, but the form is different.

The cooling and phase change is commonly seen as steam (gas) cooling to water (liquid) then further cooled to ice (solid).

10. Consolidation of solids develop into celestial objects

Galaxies, stars, and planets developed in myriad forms, infinite number, and diverse times. Stars, because of temperature, are a collection of gasses held together by gravity and inertial motion with electrical-magnetic potential.

11. The sun formed

The sun is formed at 2/3 of the time between the cosmic impulse to current time. That number is extremely fascinating. Since our sun initiation is a definable event, the time interval will critically allow definition of the time curve for cosmic development.

By the NASA account, our sun began shining at 9 billion years. Other traditions use a very different time period. Regardless, both analyses have the sun developing two-thirds of the time between the impulse and the appearance of humans.

Continuing the development of the cosmos, multiple stars form into galaxies. In a common model, a galaxy is a swirling pinwheel of space matter, which almost appears to be 2-dimensional. The diameter of the pinwheel is huge, but very thin. According to the NASA hypothesis, at the center of the pinwheel galaxy is a black hole. The black hole contains dark matter. Dark is used to indicate the material is undefined and unknown. Sure sounds like the beginning.

An alternate mathematical model of the universe is somewhat like an expanding balloon. [see our paper references in the bibliography] The shape is a three-dimensional surface of an expanding complex spheroid. Under that condition, only matter exists without black material.

Stars are the last cosmic event.

12. Plant life appears.

The remaining development is earth specific. As yet, we have not identified life forms on other celestial bodies. This phase in development is derived from terrestrial sources. NASA does research in this area as astro-biology.

Plants are the first form of life and only exhibit birth, living, and demise. No mental component is observed.

13. Water creatures and winged critters come into existence.

Life is a complex, tenuous mechanism which exists in a very narrow set of conditions. Because of this, a recent model proposes life on earth may be the first in the universe.

14. Mammals then humans appear

The warm-blooded creatures have been around for a very short term compared to any cosmic reckoning. By most accounting, if the time for mammalian existence were subtracted from the cosmic time, the cosmic time would be largely unchanged.

15. Humans are unique.

Plants have physical life. Animals have instinct, emotions, or life-giving soul. Humans are the only kind with reasoning intelligence or spirit.

16. Elapsed time to present stable or rest state.

How long was the process of cosmic implementation? Any continuous curve can be described by three points. Our interest is the curve related to human involvement from the time of the cosmic impulse.

The first consistent point is at the initialization, which is time at zero. The second point is at sun visibility, which is 2/3 of elapsed cosmic creation. The third point is recorded history, which is a rest or relatively stable state compared to cosmic progression.

The cosmic background temperature has decreased from very high temperature at the impulse to the present space temperature of 2.73 Kelvin (-270.27 degrees Celsius), which is very, very cold.

Energy losses decrease as the temperature approaches absolute zero (0 Kelvin, -273.15° Celsius, –459.67° Fahrenheit). That is one definition of the Third Law. If the temperature were achievable, then all energy transfer would stop.

> *The order and consistency of natural law leads to an inescapable conclusion which defies randomness.*

Chapter 4
Where We Are

Wow! The sequence for the formation of all within the cosmos is obviously very involved. Nevertheless, kids can understand the process, if not the details, in answer to the question, 'where did we come from?'

1. Before the beginning of the cosmos, nothing observable existed.
2. Time begins.
3. Matter defined.
4. Singularity
5. Space, the final frontier
6. Total darkness
7. Energy equation. The first epoch is complete.
8. Liquids precipitate.
9. Solids coalesce.
10. Consolidation of solids develop into celestial objects.
11. The sun formed.
12. Plant life appears.
13. Water creatures and winged critters come into existence.
14. Mammals appear.
15. Humans are unique.
16. Elapsed time to present relatively stable or rest state.

The energy of the cosmos began instantly from nothing, expanded in an impulse or inflation, then diffused in an exponential decay, as shown in the following diagram. All the energy still exists, but the form has changed and spread to configure the universe. The present condition is a very low space temperature which now is changing very slowly.

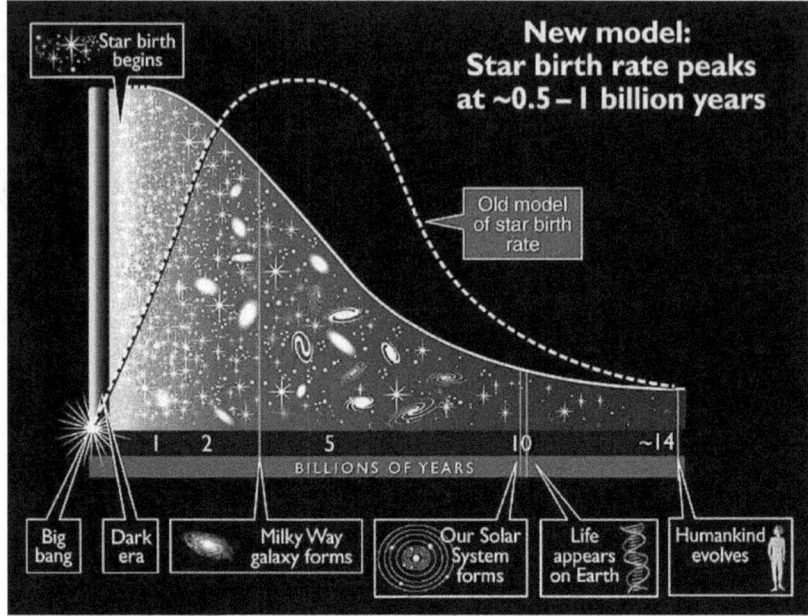

https://cdn.spacetelescope.org/archives/images/screen/opo0202b.jpg, recorded 09/27/2017.

Although the time periods involved are subject to argument by various vested interests, the sequence of the process is largely accepted. If that is the case, then reasonable minds should be interested in developing understanding of the beginning. Emotions and feelings will do little for any discussion.

Concurrence on the sequence and states of existence should lead to some version of consensus on the process of development. Remember, science is a very specific process defined by seven steps. The following steps are one way to define the scientific method.

1. Recognize a problem exists
2. Define problem to be solved
3. Gather data
4. Analyze data
5. Develop hypothesis

6. Test hypothesis by comparing to known information.
7. Repeat steps from 3, until final hypothesis.

NASA observes the dynamic nature of science. Strangely in our society, the word scientific is used to proclaim absolute truth. In reality, proper science is constantly in a state of analysis for changing accepted facts.

> As we discover more and more about the origins of our early universe, we should realize that our present theories must be continually tested and modified because new theories frequently arise as we learn more through our observations. That is why most physicists and astronomers today are so inclined to accept the Big Bang Theory as the most plausible explanation for the origin of the universe. It puts together so many of the pieces of how the universe came into being, and seems to correct so many of the flaws found in previous theories.
>
> https://lambda.gsfc.NASA.gov/product/suborbit/POLAR/cmb.physics.wisc.edu/tutorial/briefhist.html, recorded 09/27/2017.

Therefore, when different data are available, science should readily abandon previous hypotheses, if a better hypothesis fits the known data and knowledge. The acceptance of new science becomes difficult if analysis is compromised by feelings and emotion rather than reason.

Emotion is a double-edged sword, cutting deeply in any direction. Frequently, scientists reject acceptance of philosophy outside their area of expertise. Hence, some scientists do not consider historical philosophy or religious tradition. Swinging from the other direction, some religious philosophers reject science which does not conform to their tradition. Both are missing an opportunity to expand their wisdom, knowledge, and understanding.

Creating the opportunity to look at cosmology from consensus, then NASA science, followed by math, then religious tradition is our purpose. We are simply trying to answer the kid's question, 'Where did we come from?'

Chapter 5
NASA

The sequence of cosmogony is illustrated with predominantly NASA quotes. One thing to observe is cosmologist and astronomers use words like 'we are not sure that', 'I think that', 'it could be that.' What is the common theme of all these phrases?

> Cosmology is the scientific study of the large scale properties of the universe as a whole. It endeavors to use the scientific method to understand the origin, evolution and ultimate fate of the entire Universe. Like any field of science, cosmology involves the formation of theories or hypotheses about the universe which make specific predictions for phenomena that can be tested with observations. Depending on the outcome of the observations, the theories will need to be abandoned, revised or extended to accommodate the data. The prevailing theory about the origin and evolution of our Universe is the so-called Big Bang theory.
>
> https://map.gsfc.NASA.gov/universe/, recorded 09/25/2017.

Several NASA sites provide significant info about the origin. Because of the numerous associations, the link will be provided immediately after the information, rather than as a reference. The format is to identify the earlier listed sequence, then quote NASA in a related forum.

1. Before the beginning of the cosmos, nothing observable existed.

    It is beyond the realm of the Big Bang Model to say what gave rise to the Big Bang. There are a number of speculative theories about this topic, but none of them make realistically testable predictions as of yet.

https://map.gsfc.NASA.gov/universe/bb_concepts.html, recorded 09/25,2017.

The theory of a Hot Big Bang is the most widely accepted hypothesis for the origin of the universe, but it still leaves many questions unanswered. How and why did the universe expand? What caused the galaxies to form? But, perhaps the most daunting question is: "What existed before the Big Bang?" These are questions that are very difficult to answer. Astronomers are like archeologists who study the fossils of the universe. They form their theories based on what they observe, and luckily Big Bang theory seems to fit rather nicely to these observations.

https://lambda.gsfc.NASA.gov/product/suborbit/POLAR/cmb.physics.wisc.edu/tutorial/bigbang.html, recorded 9/27/2017

So how did it all start? A very good question, and one that is highly debated. Most people agree that the universe started very small and very dense and underwent an initial inflation that lasted a infinitesimal fraction of a second (thus the universe expanded much faster than the speed of light (and you thought nothing could move faster than the speed of light...).

https://lambda.gsfc.NASA.gov/product/suborbit/POLAR/cmb.physics.wisc.edu/tutorial/bigbang.html, recorded 9/27/2017

The Big Bang theory has a problem, say scientists. It can't go from a tiny ball of energy to the universe we see today without some help. It needs an adjustment called inflation.

https://cosmictimes.gsfc.NASA.gov/online_edition/1993Cosmic/inflation.html, recorded 09/27/2017.

2. Time begins

The key idea behind inflationary theory is the notion that the universe underwent a period of accelerated expansion during the first 10^{-34} seconds (0.00000000000000000000000000000001 seconds).

https://map.gsfc.NASA.gov/site/faq.html, recorded 09/27/2017.

With the sudden expansion of a pinhead size portion of the universe in a fraction of a second, random quantum fluctuations inflated rapidly from the tiny quantum world to a macroscopic landscape of astronomical proportions. Why do we believe this?

https://map.gsfc.NASA.gov/universe/uni_life.html, recorded 09/25/2017.

Notice one very significant statement. The concepts presented are a belief system. Although developed by the scientific method, the ideas are what we think is true. **What is the difference between this belief and any religious tradition?**

The following is an excellent synopsis of the beginning. The first paragraph and the last sentence are particularly intriguing. The universe began instantly at a point from a cause which is not understood.

So the 1929 discovery by Edwin Hubble that the Universe is in fact expanding at enormous speed was revolutionary. Hubble noted that galaxies outside our own Milky Way were all moving away from us, each at a speed proportional to its distance from us. He quickly realized what this meant that there must have been an instant in time (now known to be about 14 billion years ago) when the entire Universe was contained in a single point in space.

According to the theories of physics, if we were to look at the Universe one second after the Big Bang, what we would see is a 10-billion degree sea of neutrons, protons, electrons, anti-electrons (positrons), photons, and neutrinos. Then, as time went on, we would see the Universe cool, the neutrons either decaying into protons and electrons or combining with protons to make deuterium (an isotope of hydrogen). As it continued to cool, it would eventually reach the temperature where electrons combined with nuclei to form neutral atoms. Before this "recombination" occurred, the Universe would have been opaque because the free electrons would have caused light (photons) to scatter the way sunlight scatters from the water droplets in clouds. But when the free electrons were absorbed to form neutral atoms, the Universe suddenly became transparent. Those same photons - the afterglow of the Big Bang known as cosmic background radiation - can be observed today.

This and other cosmological problems could be solved, however, if there had been a very short period immediately after the Big Bang where the Universe experienced an incredible burst of expansion called "inflation." For this inflation to have taken place, the Universe at the time of the Big Bang must have been filled with an unstable form of energy whose nature is not yet known.

But supernovae observations showed that the expansion of the Universe, rather than slowing, is accelerating. Something, not like matter and not like ordinary energy, is pushing the galaxies apart. This "stuff" has been dubbed dark energy, but to give it a name is not to understand it.

http://science.NASA.gov/astrophysics/focus-areas/what-powered-the-big-bang/, recorded 3/5/2013.

3. Matter defined

This paragraph is repeated because it includes matter definition.

According to the theories of physics, if we were to look at the Universe one second after the Big Bang, what we would see is a 10-billion degree sea of neutrons, protons, electrons, anti-electrons (positrons), photons, and neutrinos. Then, as time went on, we would see the Universe cool, the neutrons either decaying into protons and electrons or combining with protons to make deuterium (an isotope of hydrogen). As it continued to cool, it would eventually reach the temperature where electrons combined with nuclei to form neutral atoms.

https://science.NASA.gov/astrophysics/focus-areas/what-powered-the-big-bang, recorded 09/25,2017.

After the Big Bang, the universe was like a hot soup of particles (i.e. protons, neutrons, and electrons). When the universe started cooling, the protons and neutrons began combining into ionized atoms of hydrogen (and eventually some helium).

https://jwst.NASA.gov/firstlight.html, recorded 09/25,2017.

4. Singularity

> The point where all that mass is trapped is called a singularity. It may be infinitely small, but its influence is enormous.
>
> https://spaceplace.NASA.gov/black-holes/en/, recorded 09/25,2017.

> For this inflation to have taken place, the Universe at the time of the Big Bang must have been filled with an unstable form of energy whose nature is not yet known.
>
> https://science.NASA.gov/astrophysics/focus-areas/what-powered-the-big-bang, recorded 09/25,2017.

For an event to occur, a cause must exist. Clearly, the cause is still unknown or unacknowledged.

5. Space, the final frontier.

> One explanation for dark energy is that it is a property of space. Albert Einstein was the first person to realize that empty space is not nothing. Space has amazing properties, many of which are just beginning to be understood. The first property that Einstein discovered is that it is possible for more space to come into existence. Then one version of Einstein's gravity theory, the version that contains a cosmological constant, makes a second prediction: "empty space" can possess its own energy. Because this energy is a property of space itself, it would not be diluted as space expands. As more space comes into existence, more of this energy-of-space would appear. As a result, this form of energy would cause the universe to expand faster and faster. Unfortunately, no one understands why the cosmological constant should even be there, much less why it would have

> *exactly the right value to cause the observed acceleration of the universe.*
>
> *We are much more certain what dark matter is not than we are what it is. First, it is dark, meaning that it is not in the form of stars and planets that we see. Observations show that there is far too little visible matter in the universe to make up the 27% required by the observations.*
>
> https://science.NASA.gov/astrophysics/focus-areas/what-is-dark-energy, recorded 09/25,2017.

> *Is it possible to travel at warp speed? Intriguingly, there are solutions to general relativity where you can warp space around you and have space move quickly towards other objects while you "surf" the rapidly moving piece of space.*
>
> https://map.gsfc.NASA.gov/site/faq.html, recorded 09/27/2017.

NASA is currently working on the first practical field test toward the possibility of faster than light travel.

> *Traveling faster than light has always been attributed to science fiction, but that all changed when Harold White and his team at NASA started to work on and tweak the Alcubierre Drive. Special relativity may hold true, but to travel faster or at the speed of light we might not need a craft that can travel at that speed. The solution might be to place a craft within a space that is moving faster than the speed of light! Therefore the craft itself does not have to travel at the speed of light from it's own type of propulsion system.*
>
> https://thesingularityeffect.wordpress.com/tag/NASA/, recorded 09/25/2017.

6. Total darkness

   Why is a powerful infrared observatory key to seeing the first stars and galaxies that formed in the universe? Why do we even want to see the first stars and galaxies that formed? One reason is... we haven't yet! The microwave COBE and WMAP satellites saw the heat signature left by the Big Bang about 380,000 years after it occurred. But at that point there were no stars and galaxies. In fact the universe was a pretty dark place.

   https://jwst.NASA.gov/firstlight.html, recorded 09/25,2017.

   One explanation for dark energy is that it is a property of space. Albert Einstein was the first person to realize that empty space is not nothing. Space has amazing properties, many of which are just beginning to be understood. The first property that Einstein discovered is that it is possible for more space to come into existence.

   https://science.NASA.gov/astrophysics/focus-areas/what-is-dark-energy, recorded 09/25,2017.

   A preliminary analysis of the Planck data suggests that this epoch, a period known as the Dark Ages that took place before the first stars and other objects ignited, lasted more than 100 million years or so longer than thought. Specifically, the Dark Ages ended 550 million years after the Big Bang that created our universe, later than previous estimates by other telescopes of 300 to 400 million years. Research is ongoing to confirm this finding.

   https://www.NASA.gov/jpl/planck-mission-explores-the-history-of-our-universe, recorded 09/25,2017.

The WMAP team found that the big bang and Inflation theories continue to ring true. The contents of the universe include 4 percent atoms (ordinary matter), 23 percent of an unknown type of dark matter, and 73 percent of a mysterious dark energy. The new measurements even shed light on the nature of the dark energy, which acts as a sort of anti-gravity.

https://www.NASA.gov/centers/goddard/news/topstory/2003/0206mapresults.html, recorded 09/25,2017.

A very interesting question is 'What are telescopes looking at?' Traditionally most of us think about the long tube with optic lens, which is used to visually peer at something in the distance. However, the resolution is only as good as the lens and eyes.

The visible light is actually an electromagnetic wave, just like a signal from a radio station. The signal is radiation from the radio station. Most modern cameras use electronics to capture and display their photographs. Similarly, telescopes use electronics to capture the light. So, in reality, a telescope is just a radio receiver detecting radiation from a distant station.

Now to address the question, 'What are telescopes looking at?' Not being facetious, but they are seeing (hearing) whatever they are tuned to, just like your radio receiver and camera.

7. Energy equation

Exactly what the universe's first light (ie. stars that fused the existing hydrogen atoms into more helium) looked like, and exactly when these first stars formed is not known. These are some of the questions JWST (James Webb Space Telescope) was designed to help us to answer.

https://jwst.NASA.gov/firstlight.html, recorded 09/27/2017.

When the first stars formed, it ended the dark ages, and started the next epoch in our universe

https://jwst.NASA.gov/firstlight.html,   recorded 09/27/2017.

The first epoch is complete. No new matter, space, or time develops, but may only change form.

8. Liquids precipitate

   The Big Bang theory predicts that the early universe was a very hot place and that as it expands, the gas within it cools. Thus the universe should be filled with radiation that is literally the remnant heat left over from the Big Bang, called the "cosmic microwave background", or CMB.

   https://map.gsfc.NASA.gov/universe/bb_tests_cmb.html, recorded 09/25,2017.

9. Solids coalesce

   The chemical elements of life were first produced in the first generation of stars after the Big Bang. We are here today because of them - and we want to better understand how that came to be! We have ideas, we have predictions, but we don't know. One way or another the first stars must have influenced our own history, beginning with stirring up everything and producing the other chemical elements besides hydrogen and helium. So if we really want to know where our atoms came from, and how the little planet Earth came to be capable of supporting life, we need to measure what happened at the beginning.

   https://jwst.NASA.gov/bigBangQandA.html,   recorded 09/27/2017.

10. Consolidation of solids develop into celestial objects.

   The grand spirals we are so familiar with (indeed including our own) were formed over the course of billions of years by several different processes, including the collisions of smaller galaxies. Giant elliptical galaxies are thought to also be formed by the process of similar-sized galaxies colliding [see videos linked at the bottom of the page], disrupting each other, and merging. In fact, it is thought that nearly all massive galaxies have undergone at least one major merger since the Universe was 6 billion years old.

   https://jwst.NASA.gov/galaxies.html, recorded 09/27/2017

   There is also more to understand about the mechanisms that cause star formation-- whether it happens internal to a galaxy or because of an interaction with another galaxy or merger.

   One thing we do know is that galaxies are still forming and assembling today.

   https://jwst.NASA.gov/galaxies.html, recorded 09/27/2017.

11. The sun formed

   Until recently, astronomers estimated that the Big Bang occurred between 12 and 14 billion years ago. To put this in perspective, the Solar System is thought to be 4.5 billion years old and humans have existed as a genus for only a few million years. ...

   Measurements by the WMAP satellite can help determine the age of the universe. The detailed structure of the cosmic microwave background fluctuations depends on the current density of the universe, the composition of the universe and its expansion rate. As of 2013, WMAP determined these parameters with an accuracy of better than than 1.5%. In turn, knowing the composition with this precision, we can estimate the age of the universe to about 0.4%: 13.77 ± 0.059 billion years!

https://map.gsfc.NASA.gov/universe/uni_age.html, recorded 09/25,2017.

Gone. Vanished. Lost.

When it comes to the early history of the solar system, planetary scientists must contend with a case of nearly systemwide amnesia.

Although the solar system formed nearly 4.6 billion years ago, researchers have a pretty good record that goes back only 3.9 billion years. Yet those first 700 million years proved critical to all that followed. That's when the planets coalesced and water and other compounds essential to life were delivered to the inner planets.

https://sservi.NASA.gov/articles/the-solar-systems-big-bang/, recorded 09/25,2017.

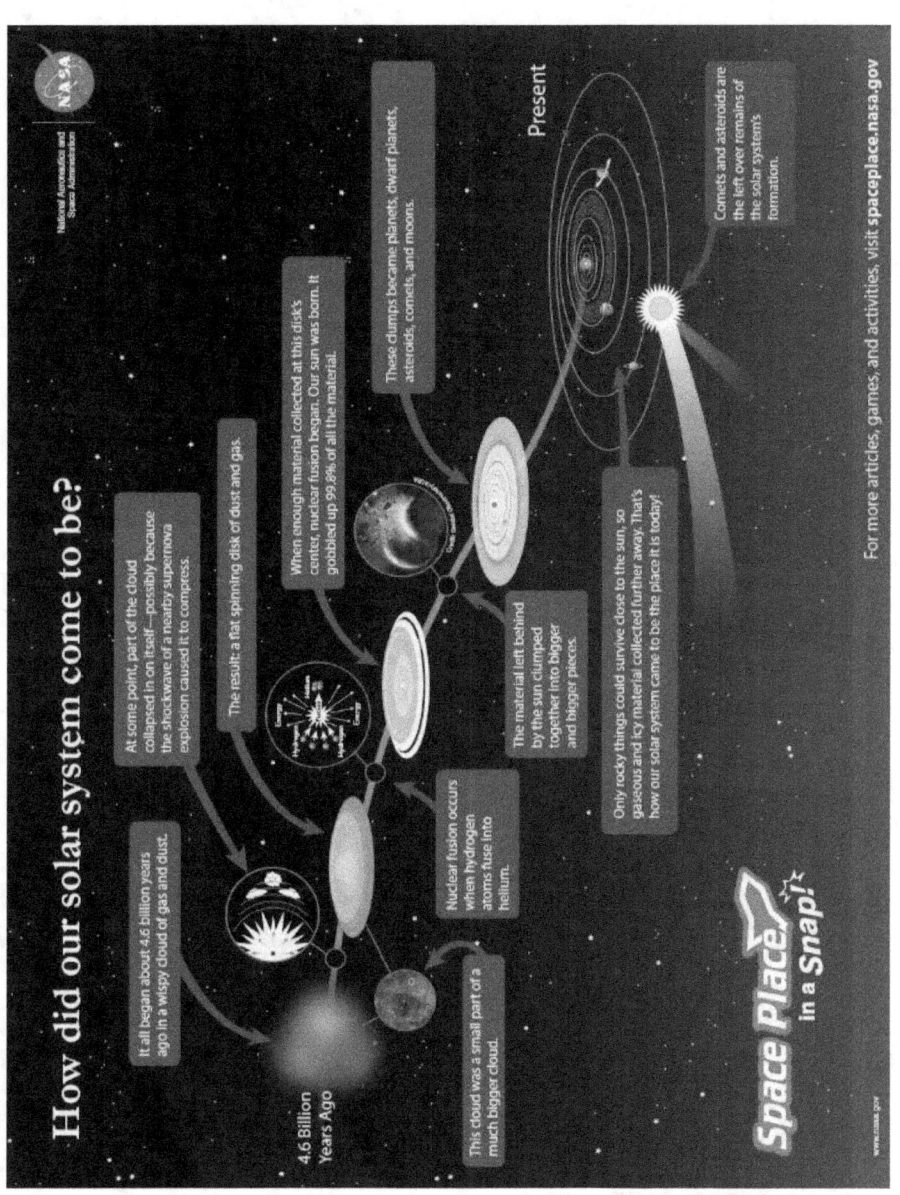

https://spaceplace.NASA.gov/review/solar-system-formation/infographic.en.png/, recorded 09/27/2017

12. Plant life appears.

First photo synthesis of plant prototypes developed roughly 1.6 billion years ago.

> *NASA exobiology researchers confirmed Earth's oceans were once rich in sulfides that would prevent advanced life forms, such as fish and mammals, from thriving. The research was funded in part by NASA's exobiology program.*
>
> *A team of scientists ... analyzed the fossilized remains of photosynthetic pigments preserved in 1.6 billion-year-old rocks from the McArthur Basin in Northern Australia.*
>
> *They found evidence of photosynthetic bacteria that require sulfides and sunlight to live. Known as purple and green sulfur bacteria because of their respective pigment colorations, these single-celled microbes can only live in environments where they simultaneously have access to sulfides and sunlight.*
>
> *The researchers also found very low amounts of the fossilized remains of algae and oxygen-producing cyanobacteria. The relative scarcity of these organisms is due to poisoning by large amounts of sulfide.*
>
> *...Team member Jochen Brocks stated "In fact, for seven-eighths of Earth's 4.5 billion-year history, there was probably little oxygen in the oceans and certainly not enough to support oxygen-breathing marine animals."*
>
> https://www.NASA.gov/home/hqnews/2005/oct/HQ_05338_toxic_seas.html, recorded 02/18/2018.

13. Water creatures and winged critters come into existence.

We generally think that NASA research is predominantly associated with space and the beginnings. NASA has a much more philosophical and almost theological charge.

> *The goal of NASA's Exobiology and Evolutionary Biology program is to understand the origin, evolution, distribution, and future of life in the Universe.*
>
> *The goal of research into the early evolution of life is to determine the nature of the most primitive organisms and the environment in which they evolved. The opportunity is taken to investigate two natural repositories of evolutionary history available on Earth: the molecular record in living organisms and the geological record.*
>
> https://astrobiology.NASA.gov/research/astrobiology-at-NASA/exobiology/, recorded 02/18/2018.

> *We don't know whether or not there is other intelligent life in the universe. There is no reason there shouldn't be. We know by our own existence that the universe is conducive to life. But there are many hurdles to overcome for intelligent life to form, and many threats to its continued existence once it does form.*
>
> https://map.gsfc.NASA.gov/universe/uni_life.html, recorded 09/25,2017.

14. Mammals appear.

Mammals and humans have been around a very short period of time on the cosmic scale.

> *The Earth and life as we know it required this heritage — you are made of star stuff! Thus, the cosmos evolved towards greater complexity in a progression that enabled the emergence of life.*

https://astrobiowalk.gsfc.NASA.gov/AstroBioPlacards.pdf, recorded 09/28/2017.

If the history of the Universe were compressed into one year with the Big Bang occurring on January 1st, Earth would be formed in mid-September, microorganisms in early October, multicellular life in early December, and Homo sapiens (humans) on December 31st.

https://astrobiowalk.gsfc.NASA.gov/AstroBioPlacards.pdf recorded 09/28/2017.

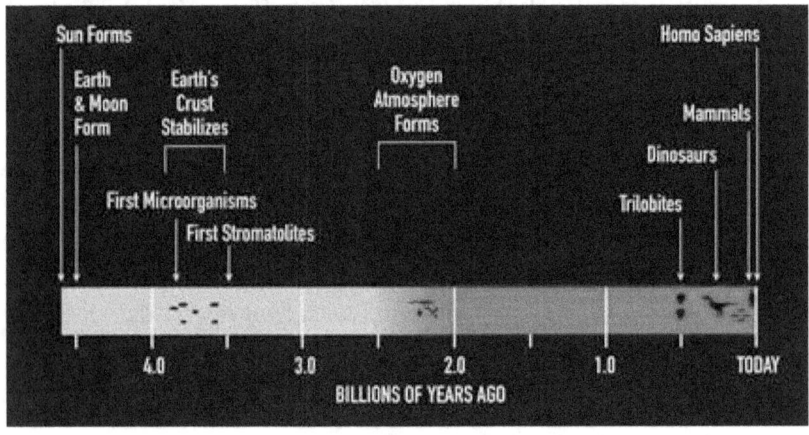

https://astrobiology.gsfc.NASA.gov/astrobiowalk/1.html, recorded 09/28/2017.

Like in other time period projections, we can largely ignore the actual numbers. The significant thing for our analysis is the sequence. The actual time numbers will be analyzed in detail in the chapter on equations. The last paragraph about relative time is an excellent illustration. Human kind is recent on the Earth's cosmic timeline.

A reviewer asked, 'Did man and dinosaurs exist at the same time? Did their times overlap? Were there fire breathing dragons during man time?' There is little definitive information one way or the other. The answer depends highly on the perspective of the respondent. Although this is not from NASA information, there are numerous other sources of information. All the history, literature, and science must be analyzed together. So, for a short answer, yes. There is ample information that dragons and dinosaurs existed with mankind. A few limited specimens seem to have existed up until a few hundred years ago.

15. Humans are unique.

    Intelligent life is how the Universe contemplates itself.

    Our Earth is unique, so far as we know, with flourishing life...

    Humanity's most buoyant characteristic is exploration and discovery. Such exploration is intrinsic to our nation and is central to the defining charter of NASA.

    https://history.NASA.gov/DPT/Science%20and%20Exploration/Science%20Priorities_04.pdf recorded 01/11/2018.

16. Elapsed time to present relatively steady, stable or rest state.

Using NASA data, the universe reached the present relatively steady process by 13.77 billion years. The last component in the sequence is human life.

>   https://map.gsfc.NASA.gov/media/060915/, recorded 09/27/2017.

The diagram below is an excellent illustration of the expanding universe from nothing, through an impulse or inflation, and progressing to recorded history.

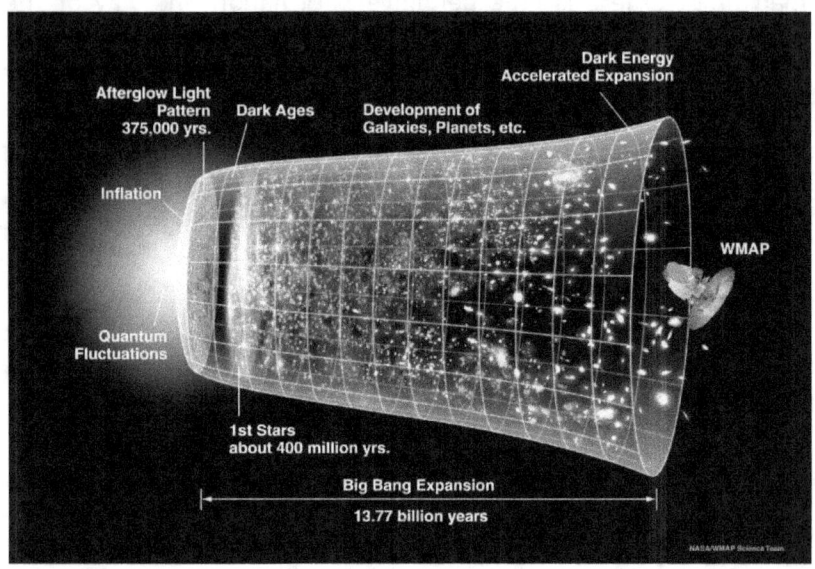

https://map.gsfc.NASA.gov/media/060915/060915_CMB_Timeline150.jpg, recorded 09/27/2017.

A representation of the evolution of the universe over 13.77 billion years. The far left depicts the earliest moment we can now probe, when a period of "inflation" produced a burst of exponential growth in the universe. (Size is depicted by the vertical extent of the grid in this graphic.) For the next several billion years, the expansion of the universe gradually slowed down as the matter in the universe pulled on itself via gravity. More recently, the expansion has begun to speed up again as the repulsive effects of dark energy have come to dominate the expansion of the universe. The afterglow

> light seen by WMAP was emitted about 375,000 years after inflation and has traversed the universe largely unimpeded since then. The conditions of earlier times are imprinted on this light; it also forms a backlight for later developments of the universe.
>
> https://map.gsfc.NASA.gov/media/060915/index.html, recorded 09 25, 2017.

Then the question is "what's next?' The exponential decay will continue until the temperature is too cold for any measurable energy.

> This is also known as the "Big Chill" or "Big Freeze" because the universe will slowly cool as it expands until eventually it is unable to sustain any life.
>
> https://map.gsfc.NASA.gov/universe/uni_fate.html, recorded 09/25/2017.

> A star like the Sun has enough fuel in its core to burn at its current brightness for approximately 9 billion years.
>
> https://map.gsfc.NASA.gov/universe/uni_age.html, recorded 09/25/2017

Based on these observations, the exponential decay continues as expected.

Chapter 6
Sidebar - A little math for non-mathematicians

The little game from an Internet circulated meme illustrates mathematic symbols in an intriguing way. From the wide-ranging discussions and varied answers, clearly numbers are a challenge to many.

This section is simply to illustrate that you can comprehend the necessary math. First thing, math is just another word for numbers. Just do not consider the terms you think you do not understand, terms like algebra! Finally, '=' is the symbol for 'is'.

This is as complicated as numbers or math becomes, in a very real sense.

$$\text{apple} = 7$$
$$\text{blackberry} = 5 + \text{apple}$$
$$\text{apple} = 1 + \text{banana}$$
$$\text{apple} + \text{blackberry} + \text{banana} = ?$$

https://misscm.com/2015/05/07/can-you-solve-this-simple-math-equation/, recorded 09/28/2017.

How many did you get?

First, in the most basic sense, the only mathematical process we can do is addition. Subtraction is addition of things you do not have, or sometimes called negative numbers. Hence subtraction is the process of 'take away'. Remember that from elementary school?

Now multiplication is just 'multiple' additions. For example, 2 times 3 is really adding 2 for 3 times or adding 2 plus 2 plus 2. But you already knew that. You can do math.

An unusual concept is multiplying a value by itself. For convenience the process of self-multiplication is called power or exponent. The symbol is the value with a small superscript for the number of repetitions. For example, 3^2 is just 3 times 3 or 9.

Math has a hierarchy. Regardless of where in the sequence, powers are done first, then multiplication and division are done next. Then addition and subtraction are done in a normal Western left to right, top to bottom pattern. We must understand the rules to obtain acceptable results.

Go back to our fruit number. To reach the final answer, take one step at a time. Then use that step answer to work the next process. Continue step by step until voila, you are there. With that, let's look at the meme.

1. So, what is the value of an apple?
2. Next what is a grape bunch worth?
3. Now what is the value of a bunch of bananas?
4. Then what is the value of *each* banana?
5. Finally, what is the result?

If you said '21', you can do the math we are going to encounter.

Drawing pictures of every word is difficult, so we use a written language. Similarly, math uses written symbols. Substitute a for apple, g for grape, b for each banana. Redraw the problem.

$a = 7$
$g = 5 + a$
$a = 1 + 3b$
$a + g + b = ?$

What is the total amount of fruit? Now one step at a time. Apple is 7, grapes are 12, three bananas are 6. Each banana is 2. So, the total amount of fruit is 21.

The sidebar is simply a minor overview of math, before we attack the equations for the cosmos.

Chapter 7
Universal equations

This section will look at the mathematical relationships which describe the steps in the cosmic sequence. The relationships are not intended to be difficult. They are simply a short-hand way that scientists use to discuss what is happening.

To aid broader understanding, the equations are separated with a *technical alert, for nerds only*. These paragraphs can be skipped. The other paragraphs will discuss the relationships in conversational language.

Even if you think you do not understand, please consider the discussion. It is critical to explaining 'where did we come from?'. Do like most of us who read technical articles outside our primary focus and experience. Read the words that make sense, skip the parts that are undue effort. Surprisingly, you will still get the idea.

All the physics and rules of nature were contained in the impulse of singularity. They became apparent at various phases of the sequence.

1. Before the beginning of the cosmos, nothing observable existed.

Time (t), matter, and space did not exist.

$$time = 0$$
$$matter = 0$$
$$space = 0$$

2. Time begins.

Time begins counting.

Three different times comprise natural science. Fixed time (t_k) is constant, so it does not change. Cyclic time (t_c) is repetitive, and is the duration of a single wavelength, cycle, vibration. Seasonal time (t_s) has a beginning, duration, and end. This is the linear time most people consider.

The time forms move from parallel to perpendicular depending on the location of the observer. In other words, if looking at a vehicle moving in front of you, it appears very slow; however, if you are looking at a cross street, any vehicle moving along the cross street appears very fast. That is a valid illustration of relativity.

Time is not a fourth dimension, but rather explains the rate an event happens. In all fundamental relationships, time is in the denominator, while the event elements are in the numerator.

3. Matter defined

Matter consists of mass, charge, and magnetism.

Mass is the component of matter experienced by the senses. Charge is the component which holds electricity. Magnetism is the component which attracts and repels from poles.

The next diagram somewhat illustrates the components, whether on the moon as shown or on the smallest atom.

The globe is the moon but is the same for a planet including the earth. First, the globe is made of matter. The most obvious is the mass, which is visible.

Second, near the top and bottom are the ends of the magnet, called the magnetic poles or polar region.

Third, as the mass with magnet turns, electric charges are built up from solar ions and static. This combination of electric charge and magnets produces weather, in the form of winds, thunderstorms and all their derivatives.

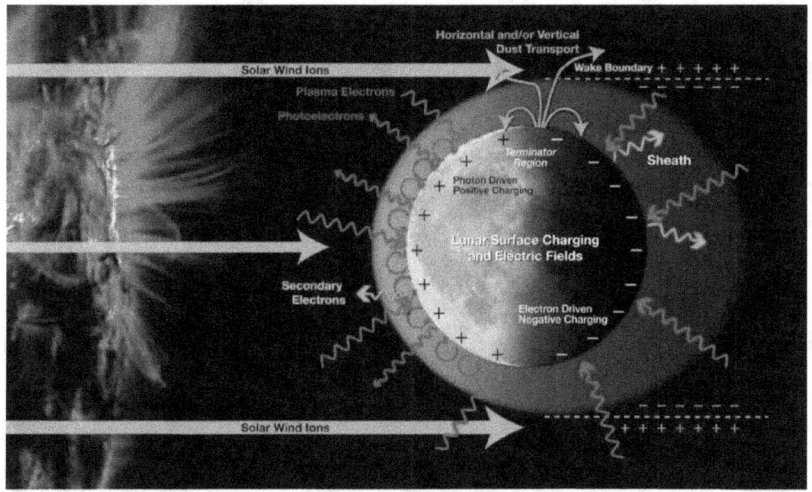

Figure 7-1: Matter Arrangement

https://www.jpl.NASA.gov/infographics/uploads/infographics/full/10689.jpg, recorded 10/05/2017.

Initially only the parts of matter existed before any consolidation into recognizable form. Then these components were combined to form elements and all known matter.

The condition of going from nothing to a peak value instantaneously is called an impulse function. An impulse is what your car feels when it hits a pothole. An impulse is the first spike on the figure we saw earlier.

An interesting question is how can the energy be decaying and the universe expanding at the same time? Initially, all energy was concentrated. As the energy spreads, it is less concentrated, and cools. The rate of reduction of energy concentration is the decay. But according to the First Law, energy is neither created nor destroyed, but only may change forms in a closed system without outside influence. Therefore, the total energy is unchanged, just the concentration is less. The reduction in energy is in the form of entropy, which is the payment to the universe for the use of energy, and is called the Second Law. The Third Law specifies the energy will go from hotter to cooler and reach stability.

Now with the energy being less concentrated, it must be dissipated some way. Consequently, the universe expands, or spreads out. So, the total energy is unchanged, but is spread over a much larger volume.

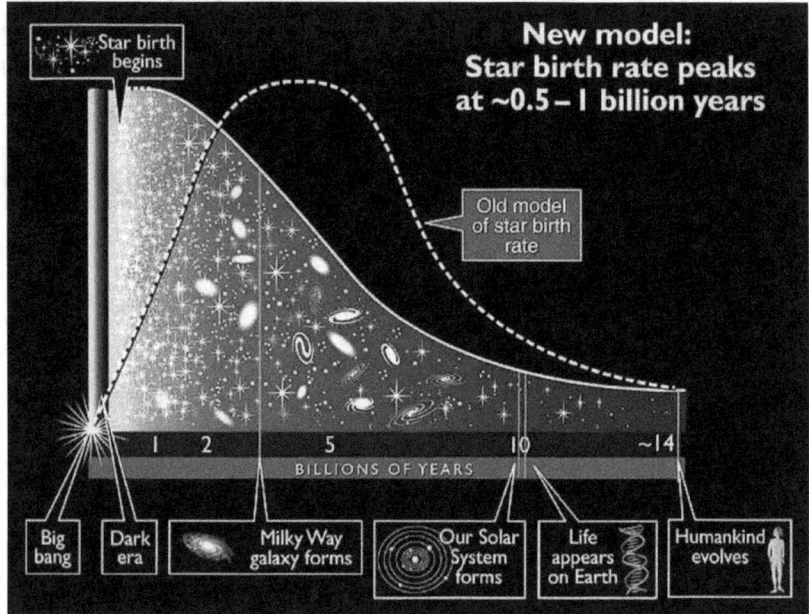

Figure 7-2: Impulse and Decay

https://cdn.spacetelescope.org/archives/images/screen/opo0202b.jpg, recorded 09/27/2017.

4. Singularity

Everything is concentrated in the impulse at the singularity, the instant after time starts.

5. Space, The final frontier

We live in a spherical type world where existence is on the surface of a ball. Volume is three dimensional. Rather than using a common cube with perpendicular arrangement, which does not exist in space, we will use a location on the surface of a sphere much like latitude (b_{rs}), longitude (d_t), and altitude, (s_r) to define space.

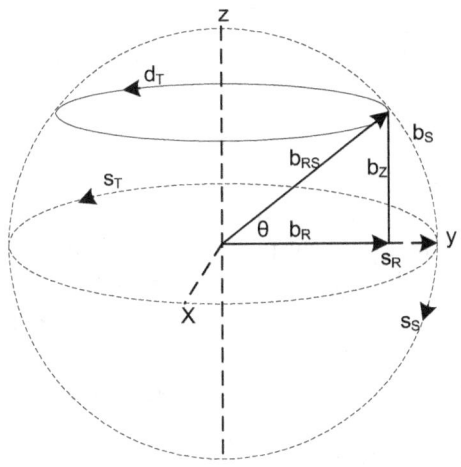

Figure 7-3: 3D globe

Now that your eyes have glazed over, let's think about the figure. This is simply a stick drawing of the globe. Everything is measured from an origin point at the nominal center of the earth. A line or axis (z) passes between the poles through the center of the globe. The equator (s_T) belts the globe. A longitude (s_S) line sets up the surface measurement east to west. A latitude line (d_T) lays out the distance north to south. The altitude (s_R) gives the height.

Because of the roundness, the concepts are considerably more complex than measuring everything from a square. But we live is this spherical world, not on a cube. So, we must look at the round measurements.

By using directional multiplication, volume (V) is the multiplication product of the three space locations like latitude, longitude, and altitude.

Technical alert: Only for nerds. The volume is a complex vector calculation where the star product (*) contains both circular or rotational and linear products. [2] The different multiplication symbols have a geometric importance.

$$V = b_{rs} * d_t \bullet s_r$$

Initially, the volume was concentrated at the impulse.

$$Volume = 0$$

Then the impulse point began expanding. The expansion can be measured and calculated. Diffusion is the rate a volume is spread in a particular direction. Volume in a direction is called volume gradient. Diffusion gives properties to space in addition to dimensions in a void. So, diffusion is simply a measure of how the universe expands. Now that was not that difficult, was it?

Technical alert: Only for nerds. The inclusion of the space vector (s_r) in both numerator and denominator is very intentional. It is the direction of movement. Among other things, it can illustrate the pancake like spiral of our galaxy observed from our planet.

$$D = V / s_r\, t_s$$

Recall back a couple of pages, noting we use time three different ways. Seasonal time has a beginning, duration and end. Notice that initially the length of seasonal time after the impulse is very short, near zero in duration. Diffusion is the volume motion divided by time. What happens to the result if we divide the volume gradient by zero? Diffusion approaches infinitely fast, hence a period called inflation. Diffusion is the name of the game commonly called big bang.

Inflation is simply another name for an impulse at the beginning.

The subsequent expansion of the universe is the diffusion of the origin point into space. Hence, the expansion when viewed from a location within the space, would appear to be away from the location. The rate of expansion when viewed would be the combination of the velocity from the viewer location and the velocity of the location from the singularity. The maximum perceived effect is when the motions are perpendicular. And that is the definition of Dr. Einstein's relativity.

The maximum velocity determined by the observer is how long it takes for the light to travel the distance to the eyes. That is defined as the speed of light. Speed of light is limited to the observer. There is no innate barrier to faster speeds in the cosmos if compared to the singularity. Note that speed of light is a transition. Lorentz's mass calculation for Einstein's very famous equation affirms you cannot exist as we know it at the speed of light. [2] It does not preclude transitioning past the speed. Just as airplanes or bullets can move past the speed of sound, producing a crack, galactic motion can move past the speed of light.

We must be careful in defining time and resulting velocity, since time perception is generally a straight-line and in one direction, but time reality is not necessarily either of these.

6. Total darkness

The morass of unconsolidated matter existed in total darkness. Darkness is the absence of light. Light is a visible evidence of energy conversion.

Energy consists of matter, space, and time. Light has matter with mass, electro- magnetism, and frequency. All the components exist, but have not been consolidated into matter or energy.

7. Energy equation

A very simple three-part relationship incorporates all known energy forms in the universe.

Previous publications have shown that everything from Einstein's mass conversion to Newtonian physics to Maxwell's suite of electro-magnetic equations are bound in these relationships. [2,4-9]

Mass-diffusion energy is simply the mass (m) multiplied by Diffusion (D) over cyclic time (t_c).

> **Technical alert:** Only for nerds.

$$E = m D / t_c$$

Electrical-magnetic energy is simply the electric charge (q) multiplied by magnetism (p) over the cyclic time (t_c).

> **Technical alert:** Only for nerds.

$$E = q p / t_c$$

Wave-constant energy is simply the natural value of Planck's universal constant (h_f) multiplied by the number of waves (w) over cyclic time (t_c).

> **Technical alert:** Only for nerds.

$$E = h_f w / t_c$$

Light results from energy conversion. Light has been described as corpuscular (mass) by Rene Descartes, electro-magnetic by James Maxwell, and as photons or discrete waves by Max Planck. All are correct, as shown in the three relationships.

The energy of light is then the energy definitions added together.

$$\frac{\textit{mass-diffusion}}{+ \textit{ electric-magnetic}} \\ \frac{+ \textit{ wave-constant}}{= \textit{light (energy)}}$$

Stated in a slightly more technical way, energy is mass multiplied by diffusion, plus charge multiplied by magnetism, plus Planck's constant multiplied by number of waves, all divided by the cyclic time. Now that is the sum of the physical energy in the universe. That was not too tough, was it?

>**Technical alert:** Only for nerds. The same information is shown in a different form.

$$light = \frac{mass \times Diffusion + electric \times magnetic + constant \times wave}{cyclic\ time}$$

Grasping this discussion is understanding the absolute working of the cosmos at its most complex levels. The geniuses Sir Isaac Newton, Rene Descartes, nor Max Planck had this much information. So, it is okay for you to say, wow! Enjoy the ride.

Energy exists. Light is. The total energy is set. Now, nothing can be created or destroyed, but may only change form. That is the definition of the First Law of Thermodynamics.

At this juncture in the cosmic sequence, all rules for the operation of the universe have been formulated and observed. Now the only thing remaining is to let the game begin.

8. Liquids precipitate

In the second phase, material changes form, but no new fundamental component of nature is involved.

The natural consequence of every impulse is a decay of the shock peak down to a condition of rest or stability. Underlying the decay down to rest is a vibration or bounce. The decay can never reach zero which is absolute. However, the value can become very, very close. Rest is finally achieved after 6 bounces times.

The figure for the impulse decay (Figure 7-4) is very well known and exists for every physical, natural phenomenon. There is nothing new or magic here.

> *The real intrigue is everything operates this same way, by a common equation or design.*

Again, reconsider the impulse from a vehicle hitting a pothole. If the vehicle is properly maintained, it is impossible to detect a vibration motion after 6 bounces, called time constants. Other vibrations may and do exist, some of which are repetitive.

> *But the final consequence of a big bang impulse is 99.8% of an impulse has dissipated to a final rest value in six (6) time-constants.* **Always. Period!**

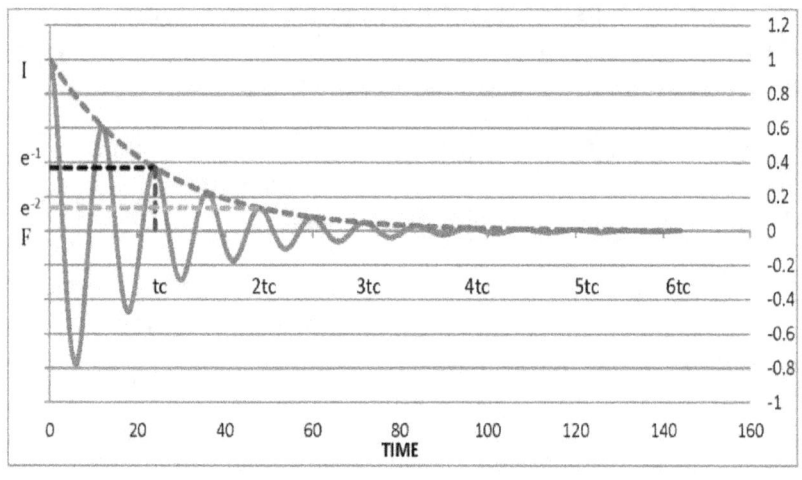

Figure 7-4: Decay function

The value of time is totally arbitrary. As we have stated numerous times, the numbers on the time is not as important as understanding the relationship, sequence, and process.

The natural decay from higher potential to lower potential illustrates two fundamental laws.

The Second Law affirms: *During every energy conversion part of the energy is returned to the universe. The return form is increasing entropy multiplied by absolute temperature.* Entropy is simply the payment to the universe for using energy.

The third Law of Thermodynamics affirms: *Natural energy flow is always from higher potential to lower.*

9. Solids coalesce

At the impulse, indescribably high hot temperatures arise. The decay from peak values results in cooling, just like every natural event. High energy (temperature) always deteriorates to lower energy (temperature) as described by the Second & Third Laws of Thermodynamics.

Gases initially exist at high temperatures. With cooling, liquids precipitate. Further cooling allows solids to take shape.

10. Consolidation of solids develop into celestial objects

Not only is the impulse peak cooling allowing material to change form, but space is expanding, and time is counting. A universe is forming.

The energy of the cosmos began instantly from nothing, expanded in an impulse or inflation, then diffused in an exponential decay, as shown in the diagram (Figure 7-2 again).

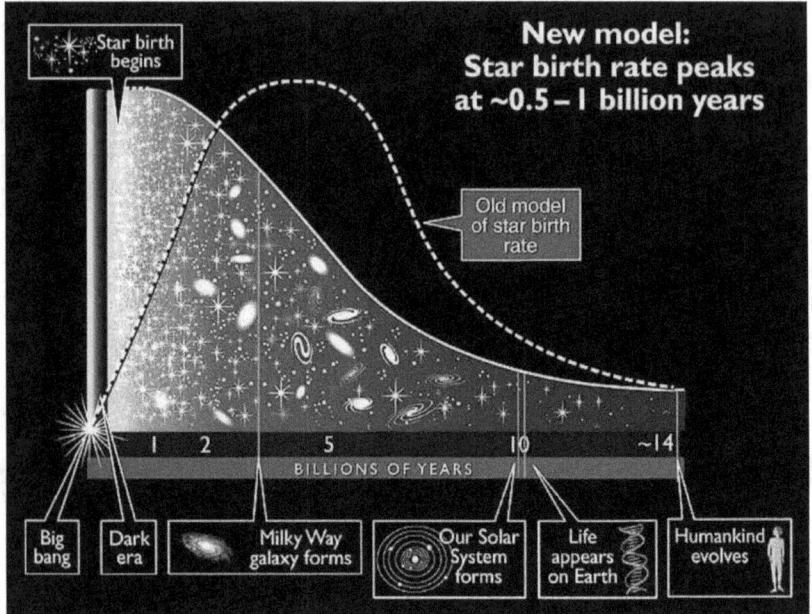

Figure 7-2 again: Impulse and Decay

https://cdn.spacetelescope.org/archives/images/screen/op o0202b.jpg, recorded 09/27/2017.

11. The sun formed

The sun is formed at 2/3 of the time from the cosmic impulse to current time. That number is extremely fascinating. Since sun initiation is a definable event, the time will critically allow definition of the exponential decay curve for cosmic development.

Three points are required to define any continuous curve, such as the exponential decay. What are three time-points which are now known? The initial impulse (I), the final rest state (F), and the time interval to the sun at 2/3 of total elapsed time have been identified.

The mathematical relationship for the sequential process is written. For the reader with less mathematical skill, simply note that a formula exists with beginning impulse and final rest values, an exponential decay (e), and other vibration waves (cosine).

Technical alert: Only for nerds.

$$E(t) = F + (1 - F)\, e^{-t/tc} \cos(\omega t + \varphi)$$

Let's camp on the exponential value for a little. This value is crucial for understanding the settling process leading up to current history.

> The concept explains in precise math the relationship between different time estimates for the creation of the universe.

Before we proceed, a new term is necessary. What is a time constant? Huh you answer. You may have little idea about the definition but you are aware it has to do with time. Although we won't go there, other equally valid questions are 'What is time?' or 'How long is time?' Time is simply a way to measure between events.

Notice the decay drawing (Figure 7-4) has values called time constants (tc). The number and effect of time constants is always the same for *every* physical system, which includes the universe.

A time constant is how long in seconds, minutes, days, or years is required for the settling to stability of the curve. The interval is always measured in the same units as the time duration called elapsed time.

Consequently, the exponent has values of -1, -2, -3, -4, -5, and -6. So, we will get the numbers for drawing the curve.

> At a time-constant beginning at 0, the energy impulse has a peak of 1.0 (e^0). So, at the beginning was everything.
>
> At a time-constant resulting in -1, the impulse has settled to a new peak of 0.367 (e^{-1}). The drop is very quick.
>
> At a time-constant resulting in -2, the impulse has settled to a lower peak of 0.135 (e^{-2}).

At a time-constant resulting in -3, the impulse has settled to a still lower peak of 0.050 (e^{-3}).

At a time-constant resulting in -4, the time of sun formation, the impulse has settled to a lower peak of 0.018 (e^{-4}). The value is less than 2%.

At a time-constant resulting in -5, the impulse has settled to a low peak of 0.007 (e^{-5}).

At a time-constant resulting in -6, the impulse has settled to a lower value of 0.002 (e^{-6}). By virtually any calculation, that is near zero. Things have settled to near rest. More periods will change very little, but are crucial.

How does this work for the solar system and universe progress?

To use sun time beginning at 2/3 of cosmic time, the time constant must be a multiple of the 3. As we just calculated, a time constant of three has not allowed the energy to settle near zero. Six is the first option which has a settling value of virtually zero.

The initial impulse (I) has a value of 1, representing everything in the cosmos. All is one. The cosmos decays, cools, and expands to a condition of relatively stable value near zero.

There are still other cyclic events occurring, but in a cosmic sense, their disturbance is minor, with 6 time-constants, the stable, rest energy settling value is less than 0.002. That means the energy is not changing very much.

By NASA's account, the sun began shining at a little past 9 billion years and recorded history began about 13.77 billion. From these numbers, a time constant is 13.77 billion / 6 time-constants. So, a time constant is 2.295 billion years or 838.25 billion days. An alternate value of time and time constant is discussed in the historical, theology section.

The value of the time constant is not as important as realizing there is a fixed reference at the time of the sun and at the relatively stable or rest condition. Then a projection can be made backward to any time, including the impulse. Star formation is the final act of the cosmic process.

>   12. Plant life appears.

The remaining phases of progression are bound to the earth. We know little about development in other worlds.

Plant life is fundamentally chemicals which propagate, grow, and demise. No evidence of mental functions exists. These organisms have a living body.

>   13. Water creatures and winged critters come into existence.

The traditional pop science attitude is other, more intelligent aliens exist and may have come to settle our world. A more recent NASA hypothesis is Earth may well be the first place of life. Based on what is known, life is a very tenuous process and can only exist under a narrow set of conditions. Disturbances either way leads to life termination.

Fish and creatures of the water appear first. Then, birds and things of the air fill the atmosphere.

>   14. Mammals appear.

Animals are obviously more complex than plants. Besides the ability for locomotion, animals have instinct, emotions, or a life-giving soul. This characteristic implies some form of brain mechanism.

>   15. Humans are unique and recorded history reach relative stability.

A requote of an earlier NASA observation will assist in human kind development.

The Earth and life as we know it required this heritage — you are made of star stuff! Thus, the cosmos evolved towards greater complexity in a progression that enabled the emergence of life.

https://astrobiowalk.gsfc.NASA.gov/AstroBioPlacards.pdf, recorded 09/28/2017.

As we have seen repeatedly, all the universe is made of the same material with a common thread, design, and equations. So, we readily agree with the first statement. However, the second statement about evolving to greater complexity is a *non-sequitur*.

The energy of the cosmos was in the original impulse, which begins the First Law (Energy is neither created or destroyed, but may only change form). Matter fused together and changed phase. The process, as has been amply shown, is an exponential decay.

The decay is reflected in the Second Law of Thermodynamics (*During every energy conversion part of the energy is returned to the universe. The return form is increasing entropy multiplied by absolute temperature*. Exponential decay leads toward stability and less complexity, which is the Third Law (*Energy naturally moves from high to low and the change approaches zero*).

In living systems, changes from the norm are rejected by the organism and the other critters of like kind. Changes cause a downward spiral, not an improvement.

Consequently, evolution does not go to greater complexity but to less complex. Moreover, to my knowledge, no transition organism has been found showing progress from one kind to another.

So, if an entity is higher order, the process is not evolution which violates all three laws of thermodynamics, but requires a new impulse of a different kind of energy.

16. Its all over.

Based on time constants, the cosmos will reach a cold state in in one more time constant. The remaining settling energy value is .0009 for time constant of -7 and .0003 for time constant of -8. Those values are virtually zero.

> The astronomers studied gas in an unnamed galaxy 7.2 billion light-years away [a redshift of 0.89].
>
> The only thing keeping this gas warm is the cosmic background radiation -- the glow left over from the Big Bang.
>
> By chance, there is another powerful galaxy, a quasar (called PKS 1830-211), lying behind the unnamed galaxy.
>
> Radio waves from this quasar come through the gas of the foreground galaxy. As they do so, the gas molecules absorb some of the energy of the radio waves. This leaves a distinctive "fingerprint" on the radio waves.
>
> From this "fingerprint" the astronomers calculated the gas's temperature. They found it to be 5.08 Kelvin (-267.92 degrees Celsius): extremely cold, but still warmer than today's Universe, which is at 2.73 Kelvin (-270.27 degrees Celsius).
>
> https://www.sciencedaily.com/releases/2013/01/130123101622.htm, recorded 09/28/2017.

The temperature change reflected in cosmic background follows the exponential decay curve as expected. With three points, the curve is defined. A defined curve permits finding any temperature at any time.

From this data, we can calculate the cosmic temperature at the time of sun formation compared to the change from the beginning. From our early decay curve (Figure 7-4) calculations, less than 2% energy change has occurred since the sun. This is further verifications that the cosmos is in a relatively stable state some call rest state.

Notice the decay curve flattens out without ever reaching the bottom. The limit on temperature is absolute zero called 0 Kelvin. The limit can never be quite reached. At that temperature all energy stops.

The universe is dead.

Life cycle

Complex specimens consist of multiple elements. Specimens include stars and celestial bodies to life forms including plants, animals, and humankind. Every complex specimen has a life cycle. The cycle includes birth, growth, duration, aging then death.

The life cycle is also called seasonal time (t_s). Notice, seasonal time is only associated with mass and diffusion. Recall, the First Law affirms energy can be neither created nor destroyed, but can only change form. Therefore, at the termination of life cycle, the mass and diffusion are the only components impacted. The energy from the mass is necessarily transformed into electric-magnetic energy and wave-constant energy or returned to the universe.

Think of the ramifications. As a human, you never cease to exist, but only change form. There is immortality after all.

Again, we find science to independently show what theologians have taught for millennia.

Pursuing the transition from mass reveals fascinating options. Throughout recorded history humans have been inferred to be a body, soul, and spirit. In contemporary parlance, humans are physical, emotional, and mental. Considering the electrical measurement of brain waves, mental is associated with electrical energy. In like fashion, emotions are associated with attraction or repelling magnetism.

> *The body (mass) has an end, but the spirit-soul (electro-magnetic) persists without time constraints.*

Chapter 8
Historical Philosophy

Thank you, philosophers and those of religious tradition for hanging on through the NASA science and the math exercise. As an astute observer, you can readily see where the discussion is going.

The oldest recorded history, philosophy, and religious document still in continual use in Western civilization is the *Torah* or *Teachings*. These first recordings were about 3500 years ago. The scribe was the adopted grandson of the Pharaoh of Egypt, with an English translated name of Moses.

As a royal scion, he had the premier education and was exposed to the court and all the knowledge of the then known world. Much of history and tradition was by stories and word of mouth. However, Moses also had access to the written history and science of humanity. He subsequently recorded a synoptic history of mankind.

The fascinating element is he succinctly recorded cosmic beginnings with uncanny accuracy. NASA has only shown a correspondence of the sequence in the past 35 or so years.

In this section, we will again describe the sequence. Similar to the NASA correlation and the previous science correlation, we will now correlate to Moses' account recorded in the English translated Genesis.

The original *Torah* was addressed to Moses' native Jewish culture. The *Teachings* have subsequently been translated into numerous languages including English. Since translations are not a word for word correlation, various translators use nuance of words. Our references will predominantly use the Complete Jewish Bible, with supplemental nuances from others.

The various quotes of the Ancient verse were obtained from http://bible.com, a phenomenal compilation of numerous translation renditions. Consequently, the spelling of some words follows the website convention, which may vary from other authors.

One consideration is important. Moses' record like other significant philosophical writings can be understood on numerous levels. There is the obvious third-grade reading as well as an intellectually subtle understanding. There is a specific understanding as well as a more generalized reading. There is a historical value as well as a theological interpretation.

The approach our analysis takes is the specific, low-level, historical, physical approach. That is a valid analytical approach. Literalists try to take each specific word and make a philosophy. Others reject historic veracity while some dismiss science. The following analysis shows validity to the science, history, and theology, which is a quite bold statement.

1. Before the beginning of the cosmos, nothing observable existed.

   We cannot discern what existed before existence.

No science record is available, so no science can be applied. Accordingly, we are left with faith or alternately conjecture, unless we can logically and philosophically come to a consensus of terms. Just because we cannot see, touch, hear, taste, or smell does not preclude developing a consistent definition.

2. Time begins.

   "In the beginning God created the heavens and the earth."

   B'resheet (Genesis) 1:1 CJB

This is the beginning statement in Moses' *Torah*.

The affirmation is at once perhaps one of the most well-known historical passages and one of the most controversial social science statements. The content is at once a religious reality and a scientific pronouncement.

The word translated heavens is from a word transliterated as 'shamayim', which comes from a root meaning aloft. The word translated heaven is otherwise translated as air or celestial location.

The word translated earth is from a word transliterated as 'eretz', which comes from a root word meaning firm. The word translated earth is otherwise translated as land, dirt, and soil. These are terms referring to **'the stuff of which things are made.'**

For the moment, leave out the deity reference. Then, with absolute translation integrity, the first affirmation has a technical form.

> "At time equal zero, 'what was before time' instigated space and matter."

The statement unequivocally is the same as NASA has declared and the science equations demonstrated. There is no difference or disagreement in the fact of an impulse start to the cosmos. There is no difference or disagreement in the initiation of matter and space in an instant.

So why is there such an adamant rejection of the reality of cosmic beginnings? The rebuff is an emotional reaction, rather than rational analysis. Social scientists are stuck in a passionate, old-school mindset, while physical scientists have accepted the reality of instant creation of the cosmos. Creation is simply defined as 'the act of producing or causing to exist.' Based on best available science at this time, that is exactly what happened at cosmic beginnings.

3. Matter defined.

In the previous chapters, we have broken the sequence into very discrete segments. Consequently, the matter-of-fact historical narrative sums up the impulse in one simple, concise sentence.

Now those of a theological tradition create quite a stir about this sentence and the next. Some of the faith-presumption is that the cosmos was created perfectly and fully blown at the beginning. If the fully-blown model is assumed, then further justifications are necessary. Three of these conflicting hypotheses are noted. I apologize if your tradition is summarized in a very short form. This is more of an identification, than a critical analysis.

The first hypothesis is that creation was perfect and complete, but the next statement affirms an unconsolidated condition. Therefore, it is argued, that creation happened, then a conflict occurred with an adversary which resulted in destruction, requiring a new act of creation. Oh, the inference surely opens many more questions.

A second creation scenario is also a perfect creation, but a long period of history occurred between the first and second sentence. The gap is necessary to explain the long cosmic age of NASA and the short days of the creative event interpreted from Moses' *Teaching*. The scenario was popular post World War II.

The third and most strongly advanced hypothesis is that the beginning occurred full-blown with the evidence of long time built into the formation. In other words, a mature earth was invoked in a young time period. The general layers of the geological column were instigated, but we have interpreted the form as being old.

The theological and religious tradition has generally rejected the NASA description of cosmic development, particularly the long time associated with the creative sequence. Furthermore, creation was an event, not a process. Any hint at a process is regarded by the 'true believer' as evolution, and evolution has not been considered as divine creation.

Unfortunately, some become so religious that they forget rational. Conversely, many are so anti-religious that they forget rational. Both camps allow their emotions to overwhelm their logic.

As a scientist and a theologian, with an earned PhD in each field, my education is conflicted. The situation forced a critical analysis of what was accepted dogma in both arenas. As it turns out, a logical, rational scrutiny shows an amazing agreement. The differences in the fields are actually just nuances. Often the differences are more a case of prejudice or unknowing than a case of understanding the foundations.

Like the word matter, 'earth' has numerous meanings. In this instance, the original 'haaretz' is a word which can reasonably be translated as matter. All the nuances of meanings are applicable in this explanation. Embedded in the word is implication of multiple different types and explanations of matter.

Matter is the whole stuff of which everything is made.

Other verses state that man is made from the matter of earth, as NASA reports everything is of the matter of the stars.

4. Singularity

A singularity is the unique point containing everything that exists. Think on that. Everything which presently comprises the entire universe at one time emanated from the single point. All the space is an expansion of that point. All matter decompressed from the point. All time is counted from the point. All physical, natural, and moral rules or laws are embedded in that point.

Physical science is based on observable data. So, by definition, science cannot describe Deity. Conversely, philosophy and theology deal with rational thought processes to explain things which are experienced, but not necessarily observable. Do you see the subtle, but very real difference in physical science and philosophy?

What would a theologian call the point from which everything emanated? God, Deity, Grand Architect of the Universe, and myriad other names, monikers, and descriptors. That is the theologians' wheelhouse.

What would a scientist call the point from which everything emanated? Singularity. That is the scientists' wheelhouse.

Why must the terms be different to describe the same point?

By no measure is the God of singularity the God of most religious practice.

Religion is predominantly a set of traditions established by authorities to control practices and actions of the adherents. Most religion is built on a variation of emotion. Since religion invokes a higher authority to achieve the goals, their God is attributed characteristics necessary to comport with the teaching of the tradition.

Why are there so many religions and so many denominations within a religion? Could it be incomplete understanding? Could it be a dedication to tradition regardless of information? Could it be they have created a God of their private understanding in conflict with evidence?

5. Space, the final frontier

Like the word space, 'heavens' has numerous meanings. In this instance, the original Hebrew word 'hashomayim' is a plural word with a definite article. The prefix 'ha' is the definite article translated to English 'the'. The suffix 'im' is the plural indicator translated to English as the suffix 's'. 'Hashomayim' can reasonably be translated as space. All the nuances of meanings are applicable in this explanation. Embedded in the plural word is implication of multiple different types and explanations of space.

Space is the whole volume in which everything exists.

Space is the expansion of the impulse point. The expansion becomes everywhere. The expansion is not from a location. There is no center. Everywhere is expanding away from all other locations. That ladies and gentlemen is the definition of infinity and omnipresence.

Space can be a measurement of distance, a void without discernible material, the region external to earth's atmosphere, and the realm of spirit beings.

What space cannot be is the residence of the Creative entity. Space expanded from the entity, so the entity cannot be inside. We can know nothing observable with the five senses about the entity or location, since it necessarily must be outside.

A cause cannot be the resultant effect.

As with each step in the sequence of beginnings, volumes can be written about the philosophy of space. A particularly fascinating aspect relates back to the energy equation, which validates numerous philosophical and religious perceptions.

Notice that space dimensions are only associated with the mass component of matter. No other matter element is constrained by space. Electric charge is unconstrained. Magnetism is unconstrained. Frequency of vibration is unconstrained, and the seven fundamental physical constants of the universe are unconstrained. [2]

ONLY mass has a space restriction on its being.

Now the story gets even more intriguing. We addressed three types of time: fixed, cyclic, and seasonal. Seasonal time is associated with a beginning, life cycle duration, and an end.

Notice that seasonal time is only associated with the mass component of matter. No other matter element is constrained by seasonal time. Electric charge is unconstrained. Magnetism is unconstrained. Frequency of vibration is unconstrained, and the seven fundamental physical constants of the universe are unconstrained. [2]

ONLY mass has a seasonal time restriction on its being.

Philosophers, theologians, and thinking individuals can have a heyday with these observations. Mass is transient. It has a beginning, life cycle, and end. But the First Law vehemently affirms, since creation 'nothing can be created or destroyed'. Is there a conflict? No. The remainder of the First Law continues 'but can only change form'.

There you have it! Based on the science of the energy equation, valid arguments can be made for spirits without mass (physical body), for a transition of life from a physical body (mass) at death to another form, and for an infinite (eternal, everlasting, perpetual) existence.

> *The pursuit of science finds itself in a strange position of validating the philosophy which theologians have argued since the beginning of recorded history.*

6. Total darkness

The next sentence from Moses' *Teachings (Torah)* continues the first step of the cosmic beginnings sequence.

> *"The earth was unformed and void, darkness was on the face of the deep, and the Spirit of God hovered over the surface of the water."*

> B'resheet (Genesis) 1:2 CJB

Matter was unformed. Matter was an unconsolidated morass. Initially, the component parts had not been incorporated into energy. The statement precisely reflects the NASA observations.

The term 'unformed and void' from the Hebrew 'tohu vavohu' is a word play also used in Isaiah 34:11. The traditional English translation interprets the catch as 'confusion and emptiness'. Other interpreters use 'formless and empty'.

An alternative reading from the Catholic Public Domain version, further enhances the image.

> *"But the earth was empty and unoccupied, and darknesses were over the face of the abyss; and so the Spirit of God was brought over the waters."*
>
> Genesis 1:2 CPDV

An obvious implication is matter did not occupy space at this interval.

The next phrase of the compound sentence is significant and the source of our title. The material and space were an abyss, an unfathomable morass of loose elements. Everything of the universe was in this concise jumble.

What about the darkness? If there is no definition of elements and compounds, no light source can exist.

Again, we see that the present NASA description of origins precisely matches the ancient writings. Fascinating stuff.

Darknesses is from another plural Hebrew word. Darkness was not an event or even a condition. Darkness is absence of light, not an opaqueness.

Since darknesses is a plural word, the implication is multiple absences of light. Multiple absences of light affirm the processes which cause light did not exist at this juncture of the sequence. The next sequence addresses light.

As a science investigation only observable processes can be analyzed. The phrase 'Spirit of God' looked over the morass moves from science to theology. Classical references specify beings as body-soul-spirit. We have previously written extensively about the classical reference of sentient beings is more appropriately stated in today's English as physical / emotional (feelings) / mental (logical, rational).

The same characteristics are attributed to the God in Moses' writing. This is not an anthropomorphic description of making God have human features. Rather, man is similar to the Grand Architect of the

Universe. In the same first chapter we have been investigating, Moses recorded the following.

> "So God created humankind in his own image; in the image of God he created him: male and female he created them."
>
> B'resheet (Genesis) 1:27 CJB

'Spirit of God' is often called the 'mind of God'. So 'Spirit of God' observing the abyss is like our saying 'He is considering' the situation. In the process of creation, the Grand Architect is going through the design steps. The beginnings were not magic or a random event. Obvious planned logic developed what are now known laws. From what is known, the cosmic beginning was also the beginning of physical, natural laws. The initializing impulse from nothing is contrary to the First Law and other rules which were not in place at the time.

Aristotle developed two words to describe the operation of the cosmos, 'energia' and 'entelecheia'. His intellect corresponds to the spirit term. [1]

> *Aristotle contemplated the One as a single entity made up of energeia. The One is the ancient source of being. The One is the source of all. The One is absolute. The One is causality. The One is. The One or the Monad is a notion of God.*
>
> *The original One emanates entelecheia, which is similar to nous by pre-Aristotle philosophers.*

7. Energy equation.

The culminating event for the first phase of beginnings is the formulation of energy. NASA reports a process like Moses.

> "Then God said, "Let there be light"; and there was light."
>
> B'resheet (Genesis) 1:3 CJB

The sentence can accurately be translated as 'let there be energy'. Light can be used interchangeably with energy. In the science section, the relationship of energy and light was defined.

Energy is the organization of matter, space, and time. With energy defined, the rules of physics and natural laws were completed. No new matter has been created since the event. The First Law of Thermodynamics affirms 'nothing can be created or destroyed, but may only change form.'

> It was a dark and stormy night at the instant of creation.

As the first act, time, space (heavens), and matter (earth) were created. These were just the unconsolidated, loose morass of pieces. Everything was dark since there was no form of energy, yet.

Then the pieces were arranged according to the laws or rules of physics and nature. One of those rules is the energy equation which describes how matter, space, and time interact.

When there is an energy conversion, light is a byproduct.

A continual light source is not necessary. For example, when a firecracker explodes, there is light, then it goes out, and we have darkness again.

So, when the laws of physics were completely defined, energy was created causing light to form.

Moses' next observation is subtle, but significant. Like many verses, there is at least a double entendre.

> "God saw that the light was good, and God divided the light from the darkness. God called the light Day, and the darkness he called Night. So there was evening, and there was morning, one day."
>
> B'resheet (Genesis) 1:4-5 CJB

Darkness is absence of light. The statement affirms the division or separation of the two. Day is defined by light, while night is defined by darkness. Then is the affirmation of darkness followed by light. First, the affirmation is the sequence of the beginning. Second, as surely as there is night, the day follows. Third, one day is a definite sequence. To this day, the Hebrew language still refers to a day as night followed by light. Western clocks have light followed by night.

Notice in the previous paragraph that three different uses were made for the word day. These are light, night+light, and era or period.

Perspective

Entire religious traditions have been disrupted by interpretation of the word day. Religious leaders have been adamant about their understanding and teaching. Barriers have developed between otherwise cogent individuals who agree on virtually everything else. Rational discussion often is impossible.

The word becomes a touchstone for a literal translation of the *Torah*. The problem is multiple literal translations are viable as we have already illustrated in the above paragraph.

> *What I know is true, because it is what I know.*
>
> ---common attitude

Two adamant extremes make a rational discussion tenuous at best. The 'anti-creation' position looks at a long evolutionary, uniformitarian model. The group rejects any religious understanding and strangely any rational scientific discussion. The position is predominantly held by political groups and social justice advocates. This is what they or their teachers were taught post World War II. So it is what they 'know'.

Unfortunately, social studies seldom involve physics or hard science. Consequently, interest or study in new scientific information escapes their information sources. The anti-creation position so rejects science that they prohibit teaching anything other than their dogma. As we discussed earlier, science - by definition - is the process of gathering new data and developing alternative hypotheses. As we have seen, NASA has created a very different model, which continues to parallel Moses' 3500-year old report.

The other extreme is 'young earth' advocates. Young earth advocates believe in an instantaneous act of beginnings. The sequence involved six literal, 24-hour days prior to recorded history. At the beginning of history, the event was complete. The long ages indicated in the geological column were built-in during the six days. The carbon-14 dating was built-in or is misinterpreted. The time for light to travel across eons and interminable space was built-in. In other words, the impulse of beginnings caused a built-in structure. Since history, the processes operate differently. In other words, the earth was created with an initial condition which appears to be old.

The short time is built on religious tradition. Therefore, a strong pathos or emotion is attached to the understanding. Significant hard science is devoted to explaining a young earth. The science interpretation validates segments of the young earth argument.

To outsiders, young-earth is blurred to mean all instantaneous genesis. The imprecision of terms makes the NASA interpretation appear as creationist, which is not the case. NASA stops at the start of time and does not have an answer for the cause. The impulse at cosmic beginnings can be both a straight-forward science observation and a straight-forward theological report as we have now observed.

So, the extremes of position are anti-creationist versus young-earth. In our discussion, we have seen there is a middle ground for cosmic beginnings. An impulse 'ex nihilo' contained all matter, space, and time. The position accurately fits both the current state of science and the traditions of rational theology.

> *Everything instantaneously came where there was nothing. Theologians have a cause for the impulse. NASA does not.*

8. Liquids precipitate.

From this point, the sequence moves in an orderly, staccato fashion.

> *"God said, "Let there be a dome in the middle of the water; let it divide the water from the water." God made the dome and divided the water under the dome from the water above the dome; that is how it was,"*
>
> B'resheet (Genesis) 1:6-7 CJB

To make sense of the phraseology requires analysis of the word translated dome. The Hebrew 'raki'a' can mean expanse, dome, firmament, or vault. The nuance is colored by the translators' world view. The dome could equally be translated a semi-sphere, which precisely matches our science model equations in the earlier chapter.

No imagination is required to use fluid instead of water in the translation.

Initially, at the elevated temperatures of the impulse, all matter was a gas which expands to fill the void which contains the fluid. With expansion and cooling, the separation of the fluid allows the denser liquid to settle while the vapor rises into the expanse. That is how it was.

We have progressed through the first two intervals of cosmic beginnings. No imagination has been required to express Moses' record into contemporary science. With the contemporary expression, the ancient record precisely correlates with the NASA record and the science equations. How awesome is that?

A side benefit of the updated translation and correlation is the added information gives theologians and philosophers an entire new realm to analyze. Moreover, this supplemental source puts to rest other spurious arguments. Similarly, scientists would be well served to understand the rudiments of theology for beginnings.

    9. Solids coalesce.

And so it goes.

> "God said, "Let the water under the sky be gathered together into one place, and let dry land appear," and that is how it was. God called the dry land Earth, the gathering together of the water he called Seas, and God saw that it was good."
>
> B'resheet (Genesis) 1:9-10 CJB

Gases fill the volume they occupy. Liquids readily flow to take the shape of their containers. Solids hold their size and shape. The same material, such as water, is a gas called water vapor at higher temperatures, a liquid called water at less temperature, and a solid called ice below a fixed temperature. The process of conversion follows the exponential decay curve of the science section.

Now Moses' record can reflect that some of the liquid under the gas is solidified.

    10. Consolidation of solids develop into celestial objects.

When the process of origins is applied, first celestial bodies were formed from matter. Then the planet has an atmosphere (gas) above, is covered by liquid called Seas and the solid matter is called earth on our planet.

Myriad types of astronomical bodies were formed and continue to develop. The cosmos is not static. Consequently, old stars and galaxies were in the early creation, as well as, newer stars and galaxies in more recent celestial time.

Again, Moses' *Teaching* parallels contemporary NASA analysis.

11. The sun formed

> "God said, "Let there be lights in the dome of the sky to divide the day from the night; let them be for signs, seasons, days and years; and let them be for lights in the dome of the sky to give light to the earth"; and that is how it was."
>
> B'resheet (Gen) 1:14-15 CJB

The statement is so concise and such an accurate definition of the sun, moon, and other body effect on our Earth's being that comments are almost superfluous. Still a few observations may solidify the parallel between the 3500-year-old statement and the present moment's science.

The Ancients have been long known to be astronomers and astute observers of the heavens. So much so that many authors of half-century ago and even social justice warriors of today dismiss the observers as superstitious, occult, or just plain ignorant.

With the tracking of more celestial bodies, NASA has further validated the periodic timing of numerous cosmic entities and the effect these have on our weather and well-being. NASA reports on these signs of astronomical events.

This year has produced multiple Atlantic hurricanes of epic size and effect. Try as social justice warriors wish, the events are the result of very natural phenomenon and not the consequence of man-made management.

The recent full eclipse across the central US was regarded with an utter sense of awe by even educated, knowledgeable scientists. The precision of the projected track of both the moon and sun is so specific that the locations where the eclipse would be greatest was known by virtually the entire society. Even school children were prepared to observe with their safe glasses, snacks, and escape from class. The event, illustrates the signs and times of the ancient statement.

12. Plant life appears.

At this point, one difference arises between Moses' recording and NASA. The Ancient report has simple plants, then seeds and trees before the formation of the Sun. Since the other points of the story match so concisely, this interchange is particularly intriguing.

So far, I have no rational explanation, but only conjecture. My tendency is to side with NASA, but the old Record has persisted so long and has been so accurate that I cannot dismiss it out of hand. My comprehension of the language hints at nothing of a different order.

A possibility has to do with many but not all of ancient astronomical perceptions. For many ancient cultures, the Earth was considered the center of the universe, but not all saw it that way. The contrarians' astronomical observations were accurate enough to forecast events with numerous examples. However, using the Earth first perception, the planet would form and begin development. Then the Sun would come, followed by the stars. Since we all are influenced to some extent by our culture, his cultural perception may have tilted Moses' writing.

In no way is this a definitive explanation of the flip in order of the events. We well know that two different observers of the precise same event will see occurrences differently. Since this is the only significant variation in the sequence, we will not become overly concerned.

13. Water creatures and winged critters come into existence.

> *"God said, "Let the water swarm with swarms of living creatures, and let birds fly above the earth in the open dome of the sky." God created the great sea creatures and every living thing that creeps, so that the water swarmed with all kinds of them, and there was every kind of winged bird; and God saw that it was good."*

B'resheet (Genesis) 1:20-21 CJB

Virtually all models accept water creatures as the first animal inhabitants of the planet. These are followed by airborne critters. The difference is how the animals arrived.

Contemporary evolutionary thought is based largely on Darwin's conjecture of natural selection published in 1859. Since he was not a geneticist, he extrapolated events. He realized that breeding such as dogs would only produce a dog. He did observe the complexity of animals, that there are simpler and more complex, and there was a commonality of some features. From this he concluded that more advanced were derived from simpler by some unknown process which resulted in an anomaly. He assumed the anomaly would be better than the previous animal.

Several problems exist with the natural selection conjecture.

    a. The first law precludes more energy and new matter.
    b. The exponential decay and Second Law preclude a progression. The law affirms the energy only decreases, therefore development of a higher order, such as life, is not possible, without external impulse.
    c. The third Law affirms static condition.
    d. Anomalies cause exclusion and are rejected in selection.
    e. No transitional form has been observed, found, or residue seen in the geological column.

Evidence based science points to discrete life forms built on a common design. The crucial question for all models is simple. What is the cause? Your answer tends to reflect your world view more than observational hard science.

Moses' model and NASA model both point to an impulse beginning with development following a precise process, which requires a design.

    14. Mammals appear.

In the cosmic sequence mammals then humans were the last significant development in the cosmos. Celestial bodies continue to birth and die. Terrestrial bodies continue to birth and die following the seasonal time script.

> *"God said, "Let the earth bring forth each kind of living creature — each kind of livestock, crawling animal and wild beast"; and that is how it was. God made each kind of wild beast, each kind of livestock and every kind of animal that crawls along the ground; and God saw that it was good.*
>
> B'resheet (Genesis) 1:24-25 CJB

15. Humans are unique.

> *Then God said, "Let us make humankind in our image, in the likeness of ourselves; and let them rule over the fish in the sea, the birds in the air, the animals, and over all the earth, and over every crawling creature that crawls on the earth." So God created humankind in his own image; in the image of God he created him: male and female he created them."*
>
> B'resheet (Genesis) 1:26-27 CJB
>
> *"Then Adonai, God, formed a person from the dust of the ground and breathed into his nostrils the breath of life, so that he became a living being."*
>
> B'resheet (Genesis) 2:7 CJB

The land animals are constructed from the earth elements. NASA stated mankind came from the stuff of the stars. Again, NASA and Moses agree.

As an extreme of the dirt idea, Greek and Roman mythology fostered the concept of Mother Earth. In their story, Mother Earth literally conceived the first mortals. Consequently, she is deified and worshipped. The thought process prevails to this day in the regressive environmental movement.

In contrast, the older and enduring record by Moses illustrates that humans are superior and have the responsibility to manage nature.

An interesting observation is each kind was unique. No transition of forms is identified. The statement is of the same form as Moses' first statement. The implication is each kind is a unique impulse event which was completed in the ancient past. The observation does not preclude the similarity of some aspects.

So far, NASA or any other science record has not identified the progression to a new kind. Wolf may be domesticated, then selective breeding can yield different breeds. But all still have canine DNA. Selective breeding produces the same kind.

Again, there is no disagreement in NASA and the *Teachings*.

The one area yet to be resolved is the attribution of the creative process. Previous comments observed such philosophy is outside the realm of observable. Subsequent discussions will further broach the philosophy and potential religious ramifications.

16. Elapsed time to present relatively stable or rest state.

The exponential decay function forecasts a time when stability or rest is reached. By now we have observed that not surprisingly, so did the *Teachings* scribed by Moses.

> *"Thus the heavens and the earth were finished, along with everything in them. On the seventh day God was finished with his work which he had made, so he rested on the seventh day from all his work which he had made."*
>
> B'resheet (Genesis) 2:1-2 CJB

As the exponential decay forecasts, stability, lack of change, or rest approaches a value of .002 in six time-intervals. The relationship heads toward zero, but will never quite get there. After the sixth time-constant, the energy level is virtually unchanged. The cosmos is at rest.

The continuing slight change will result eventually in decay to very cold and termination of current life forms by the eighth constant. That is a very, very long time by any method of time reconciliation.

A little about ages.

Moses' records some rather extraordinary ages for some of his earlier historical lineage. Various authors and scholars rationalize these periods or dismiss them as symbolic. Without prejudice about the veracity of the length of times, let's look at another perspective. After all, that is the scientific method.

> "In all, Adam lived 930 years, and then he died.
>
> In all, Shet lived 912 years; then he died.
>
> In all, Enosh lived 905 years; then he died.
>
> In all, Kenan lived 910 years; then he died.
>
> In all, Mahalal'el lived 895 years; then he died.
>
> In all, Yered lived 962 years; then he died.
>
> In all, Hanokh lived 365 years. Hanokh walked with God, and then he wasn't there, because God took him.
>
> In all, Metushelach lived 969 years; then he died.
>
> In all, Lemekh lived 777 years; then he died."
>
> B'resheet (Genesis) 5:5-31 CJB
>
> "In all, Noach lived 950 years; then he died."
>
> B'resheet (Genesis) 9:29 CJB

The time period is approaching 1000 years for each person. Interestingly, the Babylonian contemporary history also records unusually long times. There are many proposed reasons for these times. [3]

Before making assumptions, look at a few more ages. The life form changes after a period called the flood.

> "ADONAI said, "My Spirit will not live in human beings forever, for they too are flesh; therefore their life span is to be 120 years.""
>
> B'resheet (Genesis) 6:3 CJB

The implication is humans 'almost immortality' gene was lost. Ages began to shorten precipitously.

> "After Arpakhshad was born, Shem lived another 500 years and had sons and daughters.
>
> After Shelach was born, Arpakhshad lived another 403 years and had sons and daughters.
>
> After 'Ever was born, Shelach lived another 403 years and had sons and daughters.
>
> After Re'u was born, Peleg lived another 209 years and had sons and daughters.
>
> After S'rug was born, Re'u lived another 207 years and had sons and daughters.
>
> After Nachor was born, S'rug lived another 200 years and had sons and daughters.
>
> After Terach was born, Nachor lived another 119 years and had sons and daughters."
>
> B'resheet (Genesis) 11:11 - 25 CJB

Consider the age for Terach's progeny, one of the pivotal figures in world history, Abraham.

> *"This is how long Avraham lived: 175 years. Then Avraham breathed his last, dying at a ripe old age, an old man full of years; and he was gathered to his people."*
>
> B'resheet (Genesis) 25:7-8 CJB

What was Moses' age when he departed the mortal coil?

> *"Moshe was 120 years old when he died, with eyes undimmed and vigor undiminished."*
>
> D'varim (Deuteronomy) 34:7 CJB

The Psalmist, philosopher-king David, records the stable age which has continued for the last 3000 years.

> *"The span of our life is seventy years, or if we are strong, eighty; yet at best it is toil and sorrow, over in a moment, and then we are gone."*
>
> Tehillim (Psalms) 90:10 CJB

From looking at these lifespans, it should be apparent that the decrease follows what you now know as the exponential decay for everything in the cosmos. Again, the math provides additional validation that Moses' reporting was phenomenally precise and consistent with now known natural laws of physics.

How could Moses record cosmic events so precisely?

> *Before you reach a conclusion, have you read the book?*

Chapter 9
What Other Scientists Say

We started with a kids' question 'Where did we come from?' Our discussion listed the sequence of cosmic formation. Next, we correlated the sequence to NASA quotes. Then we identified the physics equations which define the operation of the cosmos. Finally, we identified the Ancient *Teachings* scribed by Moses which correlate to the sequence.

The intriguing result is all the approaches agree on the sequence beginning with nothing.

1. Before the beginning of the cosmos, nothing observable existed.

Only one major area remains to be addressed – the issue of cause. Consequently, let us look at the status of knowing the cause for the impulse of cosmic beginnings.

- Moses was trained as a Prince in Egypt.
- How did he know the sequence 3500 years ago?
- Physics recognizes the precise sequence.
- Physics also expects a precipitating cause for every event.
- *Since sciences can know nothing outside of the observable, physics cannot define the Cause.*

Therefore, we must turn to other sources and thought processes. These include scientists, philosophers, and theologians. The following excerpt is from our book, *Belief Tendencies - the Intersection of Science, Philosophy, and Theology*. [1]

Current State of Science.

Albert Einstein (1879 – 1955) was born in the German Empire and expired in Princeton, New Jersey, United States. No discussion of physics would be complete without reference to Professor Einstein, the most revered scientist of modern history and a true genius. His Theory of Relativity added changing time analysis to Newton's fixed physics. [1, 2]

In Jammer's *Einstein and Religion*, Einstein is affirmed to clarify he was neither an atheist nor pantheist. [3] An atheist believes in no god. A pantheist believes nature is identical with deity, everything comprises God. Rejection of the alternatives leaves the position where the professor was a deist or theist, since the genius believed in an actual God. Since he did not believe God was personally involved, Dr. Einstein was a deist, like so many intellectuals who reject organized religion limitations. Although he frequently quoted Spinoza, Dr. Einstein stated he knew little of Spinoza, perhaps a tongue in cheek commentary. Dr. Einstein did affirm he believed in the God of Spinoza.

Our book, *Separatists, Spinoza, & Scientists* addresses the God idea of Spinoza.

Einstein thought science shows a "Superior Mind", "Illimitable Superior Spirit", and a "Superior Reasoning Force". [4] Einstein is firmly in the camp of the reasoning rationalist tendency.

Steven Hawking (1942 – 2018) was a physicist at Oxford University and an avowed atheist. He with George Ellis and Roger Penrose published papers in 1968 and following, in which they extended Einstein's Theory of General Relativity to include measurements of time and space. [5, 6] From their measurements, they determined time and space had a finite beginning which corresponded to the origin of matter and energy. [7] The defined origin was a singularity from which everything emanated. The single source was not in space and time but space and time came from the singularity. Up to the emanating point and instant nothing existed.

> *I thought I had left the question of the existence of a Supreme Being completely open in my article. It would be perfectly consistent with all we know to say that there was a Being who was responsible for the laws of physics. However, I think it could be misleading to call such a Being 'God,' because this term is normally understood to have personal connotations which are not present in the laws of physics.*
>
> ---Steven Hawking, "The Edge of Space-Time," *American Scientist* 72, 1984, pp 355-359.

Does the statement by Hawking sound like an atheist?

Anthony Flew (1923 – 2010) was born in London and passed in Reading, England. Dr. Flew was a distinguished philosopher and professor, who was a notorious, leading atheist debater for most of his life. Dr. Flew is noted for often paraphrasing Socrates with the statement, "following the argument no matter where it leads."

Dr. Flew later wrote the book *There is a God: How the World's Most Notorious Atheist Changed His Mind.* [8] In his confession, Professor Flew wrote about the DNA structure and language which looks like "the work of intelligence." (p75) He also wrote:

> *I now believe that the universe was brought into existence by an infinite Intelligence. I believe that this universe's intricate laws manifest what scientists have called the Mind of God. I believe that life and reproduction originate in a divine Source. (p. 88)*

In an interview, Professor Flew stated:

> *Well, I don't believe in the God of any revelatory system, although I am open to that. But it seems to me that the case for an Aristotelian God who has the characteristics of power and also intelligence, is now much stronger than it ever was before. And it was from Aristotle that Aquinas drew the*

> *materials for producing his five ways of, hopefully, proving the existence of his God. Aquinas took them, reasonably enough, to prove, if they proved anything, the existence of the God of the Christian revelation. But Aristotle himself never produced a definition of the word "God," which is a curious fact. But this concept still led to the basic outline of the five ways. It seems to me, that from the existence of Aristotle's God, you can't infer anything about human behavior. So what Aristotle had to say about justice (justice, of course, as conceived by the Founding Fathers of the American republic as opposed to the "social" justice of John Rawls was very much a human idea, and he thought that this idea of justice was what ought to govern the behavior of individual human beings in their relations with others. [9]*

Professor Flew also opened the possibility of a more theistic view.

> *Yes. I am open to it, but not enthusiastic about potential revelation from God. On the positive side, for example, I am very much impressed with physicist Gerald Schroeder's comments on Genesis 1. [10] That this biblical account might be scientifically accurate raises the possibility that it is revelation.*

Professor Flew followed the evidence of a cosmological argument to move his philosophy from arch atheist to a reasoning rational deist who was tending to a perceptive maverick.

NASA Scientist

The theological implications are obvious to even the least scientific mind. But we will hear from one of the greater scientific minds.

Dr. Robert Jastrow (1925 – 2008) was a theoretical physicist and astronomer for NASA, who claimed he was agnostic. Dr. Jastrow wrote *God and the Astronomers* in which he made the following comments. [12]

> "Theologians generally are delighted with the proof that the universe had a beginning, but astronomers are curiously upset. It turns out that the scientist behaves the way the rest of us do when our beliefs are in conflict with the evidence." (p 16.)

> "For the scientist who has lived by his faith in the power of reason, the story ends like a bad dream. He has scaled the mountains of ignorance; he is about to conquer the highest peak; as he pulls himself over the final rock, he is greeted by a band of theologians who have been sitting there for centuries."

In a magazine interview, Dr. Jastrow asserted:

> "Astronomers now find they have painted themselves into a corner because they have proven, by their own methods, that the world began abruptly in an act of creation to which you can trace the seeds of every star, every planet, every living thing in this cosmos and on the earth. And they have found that all this happened as a product of forces they cannot hope to discover. That there are what I or anyone would call supernatural forces at work is now, I think, a scientifically proven fact." [13]

References

1. Albert Einstein, *Relativity, the Special and the General Theory,* Crown Publishers, New York, 1961.
2. Albert Einstein, *The Meaning of Relativity, Fifth Ed,* Princeton University Press, NJ, 1988.
3. Max Jammer, *Einstein and Religion,* Princeton University Press, 1999.

4. Denyse O'Leary, "Antony Flew, There is a God, Review," 01/01/08, http://www.errantskeptics.org/FlewAReview.htm, recorded 3/25/13
5. Steven W. Hawking & George F.R. Ellis, "The Cosmic Black-Body Radiation and the Existence of Singularities in our Universe," Astrophysical Journal, 152, (1968) pp. 25-36.
6. Steven W. Hawking & Roger Penrose, "The Singularities of Gravitational Collapse and Cosmology," Proceedings of the Royal Society of London, series A, 314 (1970) pp. 529-548.
7. Mark Eastman & Chuck Missler, The Creator: Beyond Time and Space, (1996) p. 11.
8. Anthony Flew with Roy Abraham Varghese, *There is a God: How the World's Most Notorious Atheist Changed His Mind*, Harper Collins, New York, 2008.
9. Gary Habermas, "Atheist Becomes Theist," http://www.theroadtoemmaus.org/RdLb/21PbAr/Apl/FlewTheist.htm, recorded 3/16/13.
10. Gerald L. Schroeder, *The Science of God: The Convergence of Scientific and Biblical Wisdom*, (New York: Broadway Books, 199
11. http://science.NASA.gov/astrophysics/focus-areas/what-powered-the-big-bang/, recorded 3/5/2013
12. Robert Jastrow, *God and the Astronomers*, 1978,
13. "A Scientist Caught Between Two Faiths: Interview with Robert Jastrow," Christianity Today, August 6, 1982.

Chapter 10
Other Ancient References

Just a few days ago a comedian entertainer made the comment, 'Just the ignorant believe that bible stuff.' Besides being in conflict with NASA's astrophysicist Dr. Robert Jastrow, this pseudo intellectual comment reminds me of a common advisory of my grandfather – 'It is better to be thought ignorant than to open your mouth and remove all doubt.' Grandpa was simply saying a wiser person does not engage in wasteful arguments where comments will reflect the ignorance of the speaker. Sage advice to live by.

As we look through history, we find the wisdom of my Grandpa and the opposite to the pseudo-intellectual comments. The wiser critically question conventional thinking, but come down on the side of valid, verified, veracity.

We just looked at thoughts of some of the greatest minds in contemporary thought. Although proclaimed atheists for their perception of some traditional definition of God, to the person all these intellects conceded there is a higher power and outside force, but they do not understand it. I will take that.

The longer I go and the more I understand, the more validity I see to the ancient *Tanakh* of historical Judaism and the *New Testament* of the Christian Bible. Why is that significant? Statistically, looking strictly at my education level, less than one-percent of one-percent has as much formal educational training. Without arrogance, that likely qualifies as well-educated. So intellectually, I feel confidant to state my opinions.

Throughout history, the great intellects have embraced these ancient records. Their understanding and explanations may be different from tradition, but their appreciation persists. To determine their philosophy, one must read and study their works, not the opinion of someone writing about them. Unfortunately, many writers about these gentlemen simply parrot what others say without intellectual pursuit into the base philosophy.

These geniuses and philosophers who had a significant insight into Deity include Sir Isaac Newton, Rene Descartes, Baruch Spinoza, Pierre-Simon Laplace, Dr. James Clerk Maxwell, and the inimitable Dr. Albert Einstein. That group is without doubt distinguished.

Now let us back up even further to see the status and intellect of the writers who have been included in the *Tanakh* and *New Testament*.

We have seen how the ancient historian-philosopher-leader Moses was raised as a scion of Pharaoh, the most powerful ruler of what we know as the then known world. Although he would have been in line to rule Egypt, instead he became the leader of the fledgling Jewish nation. Moses had the best education possible in the ancient world. His writing reflects this intellect. As we have seen, his recordings 3500 years ago have been validated by the highest level of contemporary cosmological science.

Some 1000 years after Moses, Jeremiah validated Moses and reflects further on the science. Jeremiah was the scion of the High Priest. In our culture, that would be similar to the Chief Justice of the Supreme Court. Jeremiah was as educated in the law as anyone of his time. His writings have persisted for 2500 years. To our point, he wrote a very significant statement. His observation will be translated into contemporary science thought or closest natural equivalence, followed by more traditional statements.

> *Wow, Sovereign Yehovah, Look around: you have made space and matter by your energy over time and your on-going work. There is nothing too hard for you.*
>
> ---Jeremiah 32:17
>
> *Adonai , God! You made heaven and earth by your great power and outstretched arm; nothing is too hard for you.*
>
> ---Yirmeyahu (Jeremiah) 32:17 CJB

What is Jeremiah saying? Like Moses he observes the formation of matter and space with energy. But, he further adds, the energy conversion process is over time. He culminates with work is on-going. Words like power, energy, and work are not interchangeable, but have very specific meanings in physics and cosmology.

The traditional translation 'great power and outstretched arm' is a Hebrew idiom *"beyad hazaqah ubizroa' netuyah"*. The phrase is attributed to Moses in Deuteronomy 26:8. Jeremiah uses it again at Jeremiah 27:5. The great philosopher-king David also used the phrase in Psalm 136:12.

Clearly these ancient writers were educated intellectuals and influential leaders. Now move forward to 2000 years ago. Paul of Tarsus was an intellectual, religious and political leader who proclaimed his education and political influence.

> *"'I am a Jew, born in Tarsus of Cilicia, but brought up in this city and trained at the feet of Gamli'el in every detail of the Torah of our forefathers. I was a zealot for God, as all of you are today. 'I persecuted to death the followers of this Way, arresting both men and women and throwing them in prison. The cohen hagadol and the whole Sanhedrin can also testify to this. Indeed, after receiving letters from them to their colleagues in Dammesek, I was on my way there in order to arrest the ones in that city too and bring them back to Yerushalayim for punishment.'*

> Acts of Emissaries of Yeshua (Acts) 22:3-5

This well-educated individual recorded the following in a letter to colleagues in Colossae, Asia Minor.

> *And he is before all things, and by him all things consist.*

> Colossians 1:17

Paul of Tarsus is affirming that the universe is made up by and held together by the Creator.

We have moved from contemporary intellectuals to early scientists, back to ancient, intellectual and influential leaders. All these affirm the same story. Therefore, a present day pseudo-intellectual who tries to project the absurdity of this historical understanding is factually displaying my grandfather's aphorism. The wise-guy has removed all doubt about his ignorance.

Chapter 11
Unique Acts

That the universe and nature's law are the result of an intelligent design has been reasonably established by accepted scientific methodology and research. Therefore, it is reasonable and logical to simply state that the cosmos came into being by a creative process.

Recently I saw a comment charging that theistic evolution should be accepted as fact. Those type observations are challenged by the scientific method. As was observed in our early discussion, there are a range of ways attempting to address the beginnings. Theistic evolution is one philosophy even recognized by NASA. But, that view is incomplete.

Our premise has been to establish a correspondence between science and ancient historic and religious tradition. That reality does not lead to theistic evolution.

Although the sequence of beginnings involves various identifiable subsequent events, the process is assuredly not evolutionary.

Evolution implies one event proceeds from a previous event. The evolutionary process necessarily implies uniformitarian development. Early discussions including observations by Edwin Hubbell in 1929 belie the uniformitarian process. Abandonment of uniformitarian thought has led to the present scientific hypothesis of a 'Big Bang' beginning.

Steady-state or evolution was rejected by cosmology about 1929. It took another half-century for the thought to spread in physics. However, soft science such as social sciences and biology still embraces the concept.

The tradition has been so embedded in education that any pronouncement away was and still is professional suicide. But the scientific method demands that alternative hypotheses be evaluated when new data is available. For the past couple of decades, the cosmological data is overwhelming against random or steady-state philosophy. The logical outgrowth is an appreciation for intelligent design.

The term evolution is often misused to reference any sequence of events. Consequently, terms like theistic evolution are used differently today from half century ago when the term was introduced. The acceptance of Deity and a cosmological process is not theistic evolution.

With the overwhelming cosmological evidence, evolution is now largely relegated to biology and to those less familiar with the hard science evidence of such a limiting view.

The beginnings involve at least seven unique creative acts which can in no way proceed from a previous circumstance. Each act is non-sequential. Each event is not the result of any other.

1. Matter begins instantaneously.
2. Time begins concurrently.
3. Space expands from nothing.
4. Natural laws are invoked to define the interplay.
5. Plant life is instilled.
6. Instinct, emotions, or life-giving soul is instilled in animals.
7. Spirit or rational intellect is instilled in humans.

In addition, a positive energy continues to cause expansion of the cosmos.

The seven creative events have three clear associations.

1. Matter, space, and time are the ingredients of physical nature.
2. Natural laws, including the laws of physics, define interaction resulting in order, structure, and energy as well as relationships.
3. Life has three facets of plant, animal, and human.

The life facet is embodied in the DNA (deoxyribonucleic acid). DNA is such an intricate link that it is mathematically impossible for the structure to have randomly or spontaneously occurred. Moreover, the DNA of each organism is unique, but contains the information of predecessors.

DNA argues against development of new kinds. To now, there has been no transition species found. Based on the mathematics, natural laws, and scientific method, an evolutionary transition species is not feasible or expected.

Even with a mapping of DNA, the spark which causes the chemistry and physics to come to life has not been identified or replicated.

Chapter 12
The Final Act

Consider the factors discussed about origins.

1. Start with the ideas from Genesis 1;
2. And with the Big Bang in precise correlation;
3. And with Dr. Jastrow's observation, the scientist is greeted by a band of theologians who have been sitting there for centuries; ... That there are what I or anyone would call supernatural forces at work is now, I think, a scientifically proven fact.

Then it is reasonable to accept the premise whereby Dr. Flew would now perceive Genesis as a revelatory account.

IMAGINE, Genesis was scribed some 3,500 years ago historically, and the account is incredibly, precisely what the epitome of space research by NASA has observed. How can such precise correlation and validation be accounted for?

Regardless of time in history, scientific propensity, or philosophical bias, the reality of Deity has been argued. Science has joined the philosophers and theologians in development of numerous correlations and consistencies. Unfortunately, inconsistencies persist.

The points of Deity in cosmogony can now be affirmatively stated.

1. A Supreme Architect exists.
2. The Supreme Architect is outside the universe.
3. The Supreme Architect created the universe instantaneously, in a fraction of a second.
4. All matter, space, time and the resulting energy emanates from the one instantaneous event and point.
5. The point of emanation is eternity, since all time and incidents are contained in the point.

6. Time provides the framework for mass-space, transcendence, and the ultra-natural realm.
7. Since the initiation, the universe operates according to absolute natural law.
8. Life is given to plants.
9. Life giving instinct, emotions or soul is given to animals.
10. Rational intellect or spirit is given to humans.
11. The universe and all of nature is expanding from positive feedback.

The Supreme Architect is still involved by providing the additional energy for the universe to grow with the benevolence of positive feedback.

The Supreme Architect does not necessarily comport with the anthropomorphic God defined by much of organized religion. The analysis of scientific philosophy reveals the Supreme correlates closely to the belief tendency of the independent reasoning of the maverick.

Dismissal of the compelling Supreme Architect concept is a rejection of rational reasoning, a rejection of physical evidence, and an acceptance of emotional bias and unbridled tradition. Decisions based on emotion or feelings is reverting to animalistic behavior without rational reasoning.

Unfortunately, most of our belief tendencies do not grasp the significance of the reality of the points of the Supreme. So, the struggles continue.

Dad, where did we come from?

We now need to finalize the answer to the kids' question.

The sequence of cosmic development is a scientific observation.

How did Moses know the sequence of cosmic development 3500 years ago?

Using the scientific method, we have shown the obvious, by all known scientific probability.

The cosmos is an intelligent design by the Grand Architect of the Universe.

Bibliography

This discussion is not a new or novel idea. Because of the nature of the work, references and links were made in the chapter involved. A few of our previous works have definitively influenced our understanding. These are generally quite technical. After all, we are making a scientific analysis of a highly philosophical topic. Most are still in print.

1. *Belief Tendencies, the Intersection of Science, Philosophy & Theology*; Marcus O. Durham, Trinity Southwest University & Realm Research, Tulsa, 2013, ISBN 978-1492313397
2. *Unified Field Theory in One Energy Equation*, Marcus O. Durham, Realm Research, Tulsa, 2012, ISBN 978-1467950701.
3. *Who Is This God?* Marcus O. Durham and Rosemary Durham, Dream Point Publishers, Tulsa, OK, 2002, 978-0971932401.
4. "Does a Unified Energy Equation Contain the Higgs Field?" Marcus O. Durham and Robert A. Durham, *IEEE Access*, Vol.1, New York, 2013, pp 505-508
5. "Engineering, Cosmogony, and Biology - Natural Law Philosophy", M. O. Durham, *Proceedings of 45th Annual Frontiers in Power Conference*, OSU, Stillwater, OK, October 2013.
6. "Dynamic Energy: Is It the Key to Electromagnetic Mass Waves?", M. O. Durham, R. A. Durham, *Proceedings of 44th Annual Frontiers in Power Conference*, OSU, Stillwater, OK, October 2012.
7. "Electromagnetics in One Equation Without Maxwell", Marcus O. Durham, *American Association for Advancement of Science - SWARM*, Tulsa, OK, April 2003.
8. "Applications Engineering Approach to Maxwell and Other Mathematically Intense Problems", Marcus O. Durham, Robert A. Durham, and Karen D. Durham, *Institute of Electrical and Electronics Engineers PCIC*, September 2002
9. "Applications Engineers Don't Do Hairy Math", Marcus O. Durham, Robert A. Durham, and Karen D. Durham, *Proceedings*

of 35th Annual Frontiers in Power Conference, OSU, Stillwater, OK, October 2002.

Annex

The science and equations presented have been a very high-level overview in an effort to keep the material more accessible to a greater segment of people. Previous publications and texts go into significantly more detail. The following is a very brief expansion of some of those highly complex concepts.

Entropy is a term used to explain energy changes, which is unknown to most people. A few definitions may help. For every energy conversion to another form, some of the energy is non-recoverable and goes to the cosmos. The unrecovered energy is called entropy. Because the original energy level has deteriorated, some authors call entropy the measure of disorder to the system. The unavailable energy is sometimes called a loss since it is unrecoverable.

My preferred definition looks positively. Entropy is our gift of energy back to the cosmos for the privilege of using energy. The universe dies the moment no more energy conversions are made.

In economics, entropy is the interest paid for use of money. In religion, entropy is the tithe paid for use of life.

With that introduction, we can address the three Laws of Thermodynamics in alternate forms. The Laws are so applicable across so many fields, that they could easily be called the first three laws of nature.

First Law

Law of Conservation (of Energy or energy components).

> *Nothing can be created or destroyed, but may only change form.*
>
> *The sum of the energy in a closed system is zero.*

Second Law

> During every energy conversion or changing of form, part of the energy is returned to the universe. The return form is increasing entropy multiplied by absolute temperature.
>
> Energy decays from higher to lower potential with a loss of energy transferred to the cosmos as entropy.
>
> The entropy of any isolated system always increases.

Third law

> Energy moves from positive, greater intensity or good to negative or less intensity.
>
> Energy loss decreases as temperature approaches absolute zero (0 Kelvin, -273.15 Celsius).
>
> The entropy of a system approaches a constant value as the temperature approaches absolute zero.

The energy decay equation contains all three laws.

$$E(t) = F + (I - F)\, e^{-t/t_c} \cos(\omega t + \varphi)$$

The initial value (I) holds all the energy from the impulse, which reflects the First Law.

The exponential with minus (e^{-t/t_c}) shows the decay from high to low, which is the Second Law.

The final value (F) is approached, which is the Third Law.

This equation is the solution or answer to every second order differential equation. Every natural system is described by a second order differential equation. So quite rationally, the relationship is the equation of everything.

The relationship between entropy (S), energy (E), and temperature (T) is illustrated.

$$dS = \partial E / T$$

Dark Matter and Energy

The known matter consisting of mass, electric charge, and magnetism represents about 4% of the known universe. The make-up of most of the universe is unknown and not understood.

To explain the unknown, NASA uses terms dark matter and dark energy. With our alternate hypothesis of expansion [2], there is legitimate science which contends there are no dark items. Nevertheless, we need to understand the discussion.

Dark matter and dark energy are called dark because they are undefined by traditional understanding. Dark matter is about 6 times greater than traditional matter or approximately 24% of the universe. Dark energy comprises about 3 times as much or approximately 72% of the cosmos. Dark matter is observed to have gravity and bend light; therefore, the matter has mass, but little if anything else. Conversely, dark energy has minimal if any mass.

Using the assumption of dark matter and energy, the terms can be reasonably defined, based on our previously published books, articles, and papers. [2, 4-9]

To begin, as a recap matter consists of mass, electric charge, and magnetism. Energy is the sum of mass-diffusion, electric-magnetic, and a wave-constant all over a cyclic time.

Now look at dark matter. With what you just read, is it really dark? No, dark matter appears to be mass without electric charge or magnetism. Dark matter appears simply to be mass which did not fuse at the cosmic initiation and which has not fused into visible matter yet. In other words, dark matter is a storage of mass for possible future fusion. Fuse simply means the coming together of components of matter.

The next unknown component of the universe is dark energy. Similarly, with what you now know, is it really dark? Well, it certainly is not light and cannot reveal light since mass is not part of the entity. So, what remains to comprise dark energy? Dark energy has three parts.

First, diffusion means spreading something. In technical terms, diffusion is space gradient over seasonal time and is the first component of dark energy. Diffusion is not an energy form since it is missing mass. Diffusion explains Einstein's opinion about properties of space being 'not nothing'. The properties of space are something real.

Next, electric-magnetic energy is an unbound, not-local energy form which has not fused.

Then, cyclic time with the waves and natural constant is stored as energy. In other words, dark energy did not fuse at the cosmic initiation and is a storage of electric-magnetics and cyclic time reserved for possible future fusion.

About the Authors

Dr. Marcus O. Durham

Dr. Marcus O. Durham is a true polymath who brings very diverse experience to his writing and lectures. He is a scientist, philosopher, theologian, researcher, author, lecturer, entrepreneur, international consultant, applied psychologist, economist, energy scholar, engineer, forensic analyst, pilot, and professor emeritus.

He is a senior principal in a scientific research lab, a principal in a forensic firm, and a principal in a natural resources company. He is a university professor emeritus of engineering, and was a seminary professor and dean.

He is a commercial pilot, is a ham radio Extra Class operator, has a commercial radiotelephone license, and is a licensed electrical contractor. He is a conservationist, who enjoys the family ranch and operating the heavy equipment.

Professional recognition includes Life Fellow-IEEE, Life Fellow-ACFE, Life Sr. Member-SPE, Certified Fire and Explosion Investigator, Certified Vehicle Fire Investigator, and Kaufmann Medal by IEEE.

Dr. Durham is acclaimed in *Who's Who of the Petroleum and Chemical Industry* and *Who's Who of American Teachers*. Honorary recognition includes Phi Kappa Phi, Tau Beta Pi, and Eta Kappa Nu.

He has published over 150 papers and articles and has authored 14 books on such diverse topics as science & engineering, economics & personal development, as well as philosophy & theology. He has traveled in over 22 countries and all 50 states.

Dr. Durham received the B.S. from Louisiana Tech University, the M.E. from The University of Tulsa, the Ph.D. in Engineering from Oklahoma State University, and the Ph.D. in Theology from Trinity Southwest University. He has other studies with numerous educational and scholarly organizations.

Rosemary Durham

Rosemary Durham is a business owner, manager, and research associate.

Professional recognition includes the following: Certified Fire & Explosion Investigator, NAFI; Certified Vehicle Fire Investigator, NAFI; Licensed FCC Amateur Radio Tech; Member, Int'l Association of Arson Investigators; Certified Color Analyst, CTI

She has co-authored two technical papers. She has co-authored three books on leadership, two books on theology, and two books for university level classes.

She is a photographer, who has analyzed the photography record for over 1000 fires and failures.

She has been active in traveling to over 15 countries and all 50 states on business and development.

She has extensive training from The Crowning Touch Institute. Her credentials are Certified Advanced Color Analyst: Introduction, Intermediate, and Advanced Color analysis and Image analysis.

Rosemary received the AB from Ayers Business College. She has additional studies at Imperial Valley College, Tulsa Community College, Oral Roberts University, Southwest Biblical Seminary and Trinity Southwest University.

Other Books

The authors have written and co-authored several books in the technical, philosophy, and applied psychology genres. Many were used in university courses. Many are still available on-line or by order from bookstores.

- *Belief Tendencies, the Intersection of Science, Philosophy, & Theology*, Marcus O. Durham
- *Unified Field Theory in One Energy Equation*, Marcus O. Durham
- *Electrical Failure Analysis for Fire and Incident Investigation*, Marcus O. Durham, Robert A. Durham, Rosemary Durham, Jason Coffin
- *Electrical Engineering in a Nutshell*, Robert A. Durham, Marcus O. Durham
- *Electrical Systems – Fundamentals for Industry*, Marcus O. Durham, Robert A. Durham, Rosemary Durham, Jason Coffin
- *Leadership & Success in Relationships & Communication*, Marcus O. Durham, Robert A. Durham, Rosemary Durham
- *Leadership & Success in Organizations, Culture, & Ethics*, Marcus O. Durham, Robert A. Durham, Rosemary Durham
- *Leadership & Success in Economics, Law, & Technology*, Marcus O. Durham, Robert A. Durham, and Rosemary Durham
- *An Intellectual's Argument About God*, Marcus O. and Rosemary Durham
- *Systems Design and the 8051*, Third Edition, Marcus O. Durham
- *Who Is This God?* Marcus O. and Rosemary Durham
- *Micro-Controllers in Systems Design*, Marcus O. Durham
- *Electrical Engineering Circuit Concepts*, Marcus O. Durham, Robert A. Durham
- *Electric Machines & Power*, Marcus O. Durham, Robert A. Durham

QED

www.ingramcontent.com/pod-product-compliance
Lightning Source LLC
Chambersburg PA
CBHW071044240526
45471CB00014B/580